118 Topics in Current Chemistry

Fortschritte der Chemischen Forschung

Managing Editor: F. L. Boschke

Oscillations in Chemical Reactions

With Contributions by
D. Gurel and O. Gurel

With 66 Figures and 11 Tables

Springer-Verlag Berlin Heidelberg GmbH
1983

This series presents critical reviews of the present position and future trends in modern chemical research. It is addressed to all research and industrial chemists who wish to keep abreast of advances in their subject.

As a rule, contributions are specially commissioned. The editors and publishers will, however, always be pleased to receive suggestions and supplementary information. Papers are accepted for "Topics in Current Chemistry" in English.

ISBN 978-3-662-15308-6   ISBN 978-3-540-38653-7 (eBook)

DOI 10.1007/978-3-540-38653-7

Library of Congress Cataloging in Publication Data. Main entry under title: Oscillations in chemical reactions.
(Topics in current chemistry = Fortschritte der chemischen Forschung; 118)
Bibliography: p. Includes index.
1. Chemical reaction, Conditions and laws of — Addresses, essays, lectures. 2. Oscillations — Addresses, essays, lectures. I. Gurel, Okan. II. Gurel D. (Demet). III. Series: Topics in current chemistry; 118.
QD1.F58 vol. 118 [QD501] 540s [541.3′9] 83-10484

© by Springer-Verlag Berlin Heidelberg 1983
Softcover reprint of the hardcover 1st edition 1983

Originally published by Springer-Verlag Berlin Heidelberg New York in 1983.

2152/3020-543210

Managing Editor:

Dr. *Friedrich L. Boschke*
Springer-Verlag, Postfach 105280, D-6900 Heidelberg 1

Editorial Board:

Table of Contents

Types of Oscillations in Chemical Reactions

Okan Gurel,[1] and Demet Gurel[2]

1 Cambridge Scientific Center, IBM Corporation, Cambridge, Mass 02142, USA
2 Department of Chemistry, New York University, New York, N.Y. 10003, USA

Table of Contents

I The Scope: Oscillations

In an experimental environment results are to be interpreted as to reveal the behavior of chemical reactions. The characteristics of this behavior determine the success in obtaining the desired results. In this context the *equilibrium states* have been the primary concern of the investigators. With the use of some mathematical formalism these states are described as *stable states* of the system under consideration.

A *mathematical model* may be constructed representing a *chemical reaction*. *Solutions* of the mathematical model must be compatible with the observed behavior of this chemical reaction. Furthermore if some other solutions would indicate possible behaviors so far unobserved, of the reaction, experiments maybe designed to experimentally observe them, thus to reinforce the validity of the mathematical model. Dynamical systems such as reactions are modelled by differential equations. The chemical equilibrium states are the *stable singular solutions* of the mathematical model consisting of a set of differential equations. Depending on the format of these equations solutions vary in a number of possible ways. In addition to these stable singular solutions periodic solutions also appear. Although there are various kinds of oscillatory behavior observed in reactions, these periodic solutions correspond to only *some* of these oscillations.

The incorporation of oscillations into the analysis of chemical reactions has attracted researchers for sometime. Many experimental observations were successful in conforming the existence of oscillations. However these observations and in some cases mathematical models form a fraction of the research on chemical reactions, thus limited knowledge has been gained. There are multiple reasons for this being short of completeness. In all cases, it points to a lack of our understanding the dimensions of the kinetics involved in chemical reactions. The design of a laboratory experiment or of a computer simulation of a specific reaction is done with a fixed set of conditions, thus the scope of the study becomes narrow. Furthermore, the theoretical models explaining the observed phenomena during this study explains only the restrained behavior of the reaction. Because either the emphasis is to find a model to match the observations only, or even if the model is rich in revealing "other" possibilities of solutions, the theoretical analysis may be carried up to explaining the small variations of the system under study.

In order to form a bridge between the laboratory (chemical) experiments and the theoretical (mathematical) models we refer to Table I. In a traditional approach, experimental chemists are concerned with Column I of Table I. As this table implies there are various types of research areas thus research interests. Chemists interested in the characteristics of reactants and products resemble mathematicians who are interested in characteristics of variables, e.g. number theorists, real and complex variables theorists, etc. Chemists who are interested in reaction mechanism thus in chemical kinetics may be compared to mathematicians interested in dynamics. Finally, chemists interested in findings resulting from the study of reactions are like mathematicians interested in critical solutions and their classifications. In chemical reactions, the equilibrium state which corresponds to the stable steady states is the expected result. However, it is recently that all interesting *solutions* both *stationary* and *oscillatory*, have been recognized as worthwhile to consider.

3

Stability of these solutions (observed states) mainly depends on the underlying dynamics (chemical kinetics), therefore for both the solutions themselves and their stability properties we have to refer to dynamics and the characteristic solutions.

In the present review, emphasis is placed on oscillations in reactions, and oscillatory solutions of corresponding models. Some preliminary information on the theory of oscillations is discussed, however only as a reference not for detailed study of the subject. It is only from this view that the current state of the art will be reviewed, and therefore many seemingly relevant discussions in the literature may not be essential for the purpose of the current review, and fall outside the scope of the present article.

Another point of interest to the reader is that except the review articles, only those contributions where the original chemical observations and advances are presented will be reviewed. The remaining articles are considered as relevant to areas outside the scope of the present review. This explanation is necessary because under the name of, e.g. Belousov-Zhabotinskii, the literature is inundated beyond reach, obviously contributing to other areas of this reaction scheme not necessarily focusing on its oscillatory solutions. Furthermore, the terminology used in the referenced sources is preserved wherever possible.

Table I. Comparison of Chemical and Mathematical Systems

	Column I Chemical Systems	Column II Mathematical Systems
Elements	• Physico-chemical unknown entities	• Variables
	• Physico-chemical known entities	• Parameters, Constants
Process	• Reaction systems (Chemical Kinetics)	• Dynamics
	• Observations of observable states	• Critical solutions in solution space
Results	• Equilibrium states	• Stable singular points
	• Oscillations (Oscillatory equilibrium states)	• Stable periodic or complex-periodic solutions

II Review Articles

In the literature there is a small number of reactions exhibiting oscillations, observed experimentally, which motivated a vast number of studies either devising a model for the reaction scheme or analyzing the small variations thereof. Although oscillatory behavior has been recognized in the past by a handful of chemists. it is recently that oscillatory behavior of chemical systems attracted considerable attention. As a result, studies carried out by various groups of researchers have been reviewed and summarized in review articles. Some of these reviews are more comprehensive than others and cover multiple examples of oscillatory reactions. A partial list of these articles is given in Table II with some annotations.

III Reactions and Models Exhibiting Oscillations

In this section we refer to various chemical reactions and proposed models known to exhibit certain oscillatory behavior. The purpose of this listing is to summarize the known systems and form a dictionary of the types of oscillatory behaviors which will be subsequently used in determining the characteristics of various oscillations and the systems exhibiting them. The order in which various systems are listed is by the original date of the study relevant to the system discussed.

A Iodate Catalyzed Decomposition of Hydrogen Peroxide (Bray-Liebhafsky Reaction)

References: Auger (1911), Bray and Caulkins (1916), Bray (1921), Bray and Liebhafsky (1931), Liebhafsky (1931-1), Liebhafsky (1931-2), Bray and Caulkins (1931).

As reported by Bray (1921) the role of hydrogen peroxide as an *oxidizing* and *reducing* agent was investigated by A. L. Caulkins and W. C. Bray starting 1916. The reactions considered were given in Bray's 1921 paper, and a credit was given by Bray to Auger for discovering these reactions in 1911.

Reaction Scheme: Auger (1911), Bray (1921)
In this example, *hydrogen peroxide*, H_2O_2, plays a dual role in,
(1) *Oxidation* of *iodine*, I_2, to iodic acid, HIO_3.
(2) *Reduction* of *iodic acid*, to iodine.
 These reactions were first given by Auger (1911) as below. However, it was in Bray's paper (1921) that oscillations in this system were recognized. Under the experimental conditions used by Bray the system yields a damped (dying out) oscillation.

$$5 H_2O_2 + I_2 = 2 HIO_3 + 4 H_2O \tag{1}$$

$$5 H_2O_2 + 2 HIO_3 = I_2 + 5 O_2 + 6 H_2O \tag{2}$$

The role of the oxidation-reduction couple in the catalytic decomposition of hydrogen peroxide,

$$H_2O_2 = H_2O + \tfrac{1}{2} O_2 \tag{3}$$

was discussed. Reaction (1) is *autocatalytic* and proceeds rapidly. Reaction (2) proceeds relatively slowly.

Mechanistic Studies and Computer Simulation of a Model:

References: Degn (1967-1), Lindblad and Degn (1967), Matsuzaki, et al. (1974), Sharma and Noyes (1974), Edelson and Noyes (1979).

For almost half a century various attempts were made to discount Bray's discovery of the oscillatory behavior of this system. However, Degn (1967) experimentally verified the oscillations, and with his colleagues (1967) argued and gave a model based on quadratically branched chain reaction giving rise to oscillations similar to the experimental results.

Table II. Review Articles on Oscillatory Chemical Systems

Reactions	Review Articles (*)															
	1	2	3	4	5	6	7	8	9	10	11	12	13	14	15	16
Bray	×		×	×					×				×			
Briggs-Rauscher				×					×							
Belousov	×		×	×					×			×				
NFK Model												×				
CSTR			×				×			×	×			×	×	
Solid Catalyzed reactions																
N₂O decomposition										×	×					
H₂ oxidation							×			×	×					
CO oxidation							×			×	×				×	
Glycolysis	×	×	×					×				×				×
Higgins M 64		×	×			×		×								×
Sel'kov M 68			×			×		×								
Peroxidase	×	×	×			×		×				×				
Yamazaki	×		—			—		×				×				
Degn	×							×								
Na₂S₂O₄ decompostion									×							
Bimolecular				×	×	×						×				
Abstract Models													×			×

Chemical Entities

	1	2	3	4	5	6	7	8	9	10	11	12	13	14	15	16
Peroxidase	×							×								
Bromate	×		×	×					×					×		
Iodate			×	×					×					×		
Concepts																
Feedback		×						×						×		
Bifurcations		×				×										×

(*) Authors and dates of review articles

1 = Degn	1972	5 = Gray	1974
2 = Gurel	1972	6 = Gurel	1975
3 = Nicolis	1973	7 = Schmitz	1975
4 = Noyes	1974	8 = Goldbeter	1976
9 = Noyes	1977	13 = Rossler	1979
10 = Sheintuch	1977	14 = Ray	1979
11 = Hlavacek	1978	15 = Gray	1979
12 = Franck	1979	16 = Gurel	1979

Further studies on the mechanism of this reaction showing oscillations have been carried out by a number of investigators, notably by H. A. Liebhafsky and his collaborators, and Noyes and his collaborators. For example, in the paper by Matsuzaki et al., the results of a computer simulation are given. Oscillations of iodide, I^-, iodine, I_2, and hypoiodous acid, HIO are shown. Although there is no specific reference to a *limit cycle*, the oscillations appear to form a limit cycle, Fig. III.1.

Furthermore, the possible bifurcations to other types of oscillatory solutions have not been considered.

In 1976, Noyes with Sharma and subsequently in 1979 with Edelson reported a detailed mechanism and calculations modeling this oscillatory reaction.

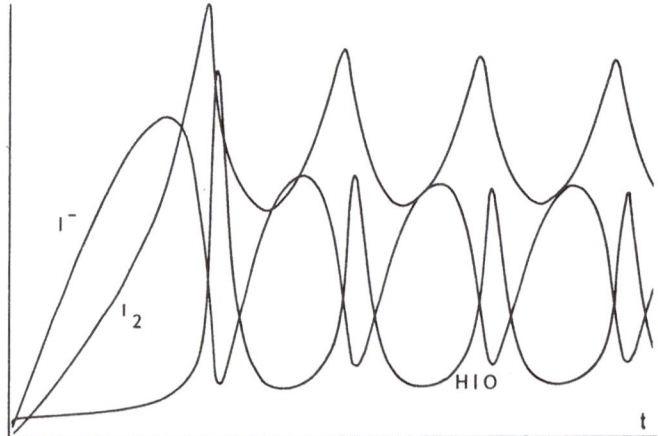

Fig. III.1. Sustained oscillations of I_2, I^-, and HIO. (After Matsuzaki et al. (1974)

B An Oscillating Iodine Clock (Briggs-Rauscher Reaction)

Reference: Briggs and Rauscher (1973)

Briggs and Rauscher discovered an oscillating reaction which is identified as an *iodine clock*. It resembles the iodate-hydrogen peroxide reaction of Bray (1921), and has some of the elements of the reaction of Belousov, see Section III.C. The chemicals involved are:

> Potassium iodate, KIO_3
> Hydrogen peroxide, H_2O_2
> Perchloric acid, $HClO_4$ (or sulfuric acid)
> Malonic acid, $CH_2(COOH)_2$
> Manganese(II) sulfate, $MnSO_4$
> Starch.

In this reaction the concentrations of iodine, I_2 and iodide ion, I^- oscillate. The resulting oscillations for iodide are shown in Figure III.2.

Substitution of 2,4-pentanedione for malonic acid results in short lived oscillations, and substitution of cerium as catalyst in place of manganese results in higher frequency oscillations.

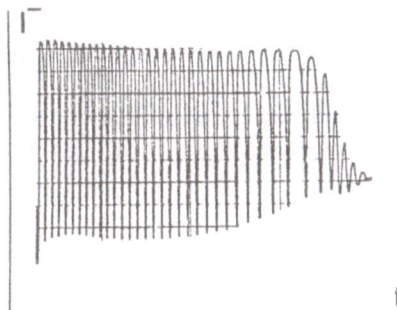

Reaction Scheme: Boissonade (1976), De Kepper et al. (1976)
As shown below in modelling this reaction various solutions were considered. For certain values of parameters it is shown that there are three singular solutions, two stable and one unstable as well as limit cycles (multiple solutions).

This reaction was subsequently studied by various groups. The model by Boissonade consists of two parts. The first part exhibits oscillations. Furthermore, the additional scheme introduces the *double oscillations*. These oscillations are the ones with periods shorter than the overall period of the reaction, and such double oscillations were previously discussed by Beusch (1972), Marek and Svobodova (1975), Zhabotinskii and his colleagues (1973), and Dynnik and Sel'kov (1975) in other systems, see Sections III.C, E and F. Boissonade (1976) gives the reaction for creating these oscillations as

$$C \xrightarrow{k_1} X$$
$$2X \xrightarrow{k_2} 2Y$$
$$Y + Z \xrightarrow{k_3} 2Z$$
$$X + Z \xrightarrow{k_4} Product$$

Differential Equations: Boissonade (1976)
Rate equations for the reaction scheme above, the differential equations, are written as the model which consists of:

$$dX/dt = k_1 C - 2k_2 X^2 - (k_4 Z + k_E) X$$
$$dY/dt = 2k_2 X^2 - (k_3 Z + k_E) Y$$
$$dZ/dt = k_0 Y + k_3 YZ - (k_4 X + k_E) Z .$$

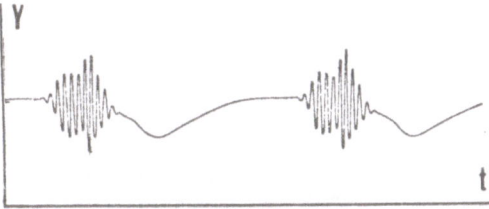

Fig. III.3. Double oscillations of the model for Briggs-Rauscher reaction (After Boissonade (1976))

Oscillations of Y are given in Fig. III.3.

In Boissonade (1976) it is also noticed that the system has either a singular point, or multiple (three) singular points and a limit cycle.

Pacault et al. (1975) reported some experimental results, while Pacault et al. + Rossi (1975) determined the domains of oscillations for elements KIO_3, $CH_2(COOH)_2$ and H_2O_2. The oscillations obtained by P. de Kepper et al. (1976) are shown in Fig. III.4.

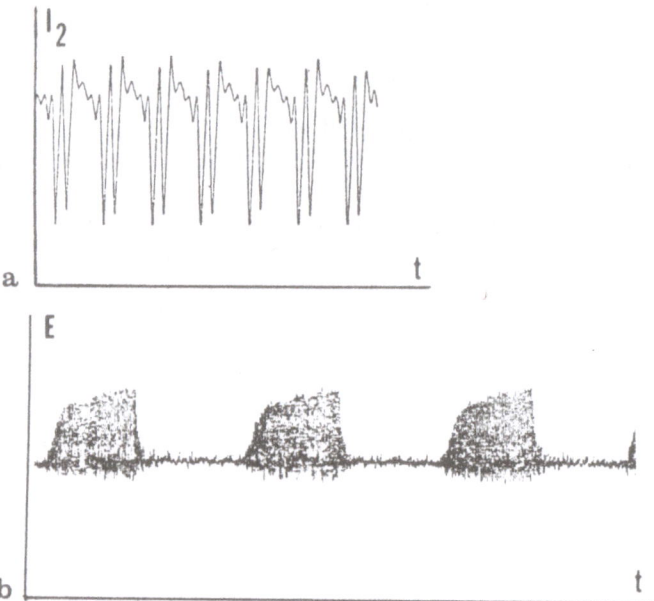

Fig. III.4. a I_2 concentration oscillations, **b** E, electrical potential oscillations. (After de Kepper 1976)

Interesting results showing both simple and complex oscillations of I_2 versus time are reported in Boissonade and de Kepper (1980). Fig. III.5 shows the simple limit cycle of the system reported by Boissonade and de Kepper (1980).

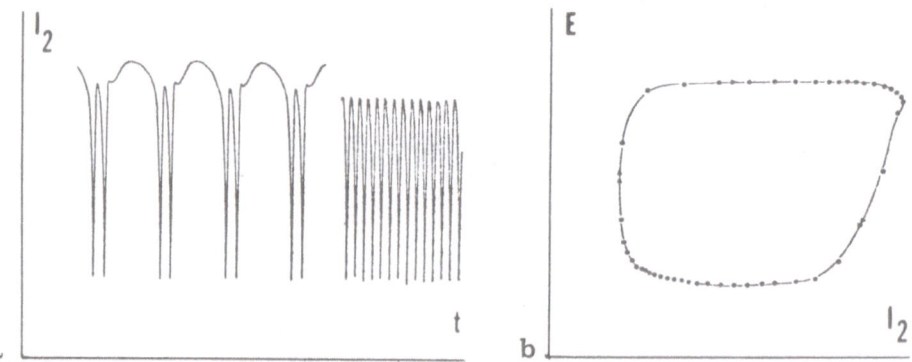

Fig. III.5. a) Simple and complex oscillations of I_2 for the Briggs-Rauscher reaction under different conditions, **b)** the limit cycle and the E, I_2-plane (After Boissonade and de Kepper (1980))

10

C Oxidation of Malonic Acid by Bromate (Belousov-Zhabotinskii Reaction)

References: Belousov (1959), Zhabotinskii (1964-1), Noyes, Field, Koros (1972), Field, Koros, Noyes (1972), Field, Noyes (1974), See also Bray and Liebhafsky (1935).

Belousov (1959) observed sustained oscillations in the ratio of concentrations of the ceric and cerous ions, Ce(IV)/Ce(III), during the cerium catalyzed *oxidation* of citric acid by bromate in aqueous sulfuric acid.

Zhabotinskii (1964-1) showed that similar oscillations are observed when malonic acid is used as the oxidizable substance.

Later it was observed that an oscillatory reaction can also be obtained when

- Citric acid is replaced by malonic acid or any other acid with an active methylenic hydrogen
- The Ce(IV)/Ce(III) couple is replaced by the Mn(III)-Mn(II) or by ferroin-ferriin couple.

Reaction Scheme: Zhabotinskii (1964-1)
Zhabotinskii proposed initially a reaction scheme to explain Belousov's reaction as

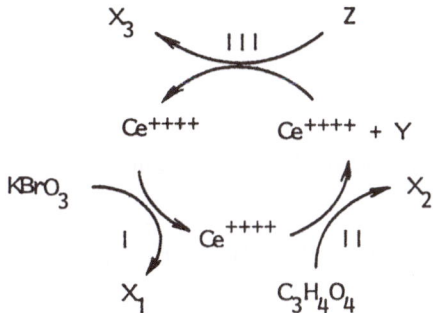

where X_1, X_2, X_3, and Y are unknown chemical compounds, Z is possibly $KBrO_3$ or an intermediate product of it. Oscillations observed of colored solution are illustrated in Fig. III.6.

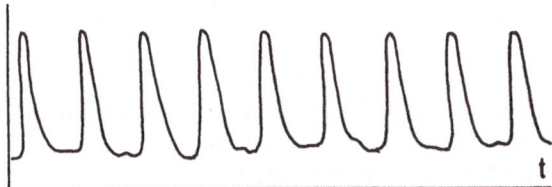

Fig. III.6. Oscillations of colored solution of Belousov reaction (After Zhabotinskii (1964-1))

Reaction Mechanism: Noyes, Field, Koros (1972), Field, Koros, Noyes (1972)
Noyes et al. (1972) gave a mechanism for the Belousov reaction. They proposed that in an acid solution of bromate, BrO_3^-, and malonic acid, $CH_2(COOH)_2$ containing sufficient amount of bromide ion, Br^-, four reactions form *Process A*. These reactions are (R1, R3, and R7).

11

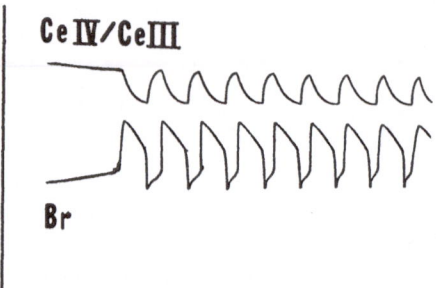

Fig. III.7. Log. (Br^-) and $\log (Ce(IV))/(Ce(III))$ (After Noyes, Field and Koros (1972))

In the presence of H^+, bromate ion reacts with bromide ion to yield bromous acid and hypobromous acid,

$$BrO_3^- + Br^- + 2\,H^+ \rightarrow HBrO_2 + HOBr \tag{R1}$$

Subsequently bromous acid reacts with bromide ion to yield also hypobromous acid,

$$HBrO_2 + Br^- + H^+ \rightarrow 2\,HOBr \tag{R2}$$

Hypobromous acid reacts with bromide ion to yield bromine,

$$HOBr + Br^- + H^+ \rightarrow Br_2 + H_2O \tag{R3}$$

Then, bromine reacts with malonic acids to yield bromomalonic acid as (= Bromination of malonic acid by bromine)

$$Br_2 + CH_2(COOH)_2 \rightarrow BrCH(COOH)_2 + Br^- + H^+ \tag{R7}$$

Therefore *Process A* is the reaction as

$$(R1) + (R2) + 3(R3) + 3(R7), \quad \text{i.e.,}$$

$$BrO_2^- + 2\,Br^- + 3\,CH_2(COOH)_2 + 3\,H^+ \rightarrow 3\,BrCH(COOH)_2$$
$$+ 3\,H_2O \tag{A}$$

bromite ion	bromide ion	malonic acid		bromomalonic acid

When bromide ion, Br^- is virtually absent, bromate ion, BrO_3^- reacts with cerium(III) and malonic acid, such that *Process B* results from four reactions.

Bromate ion reacts with bromous acid to yield radical $BrO_2\cdot$

$$BrO_3^- + HBrO_2 + H^+ \rightarrow 2\,BrO_2\cdot + H_2O \tag{R4}$$

Then BrO_2 reacts with cerium(III) ion to yield bromous acid and cerium(IV) ion (= cerium oxidation)

$$BrO_2\cdot + Ce^{3+} + H^+ \rightarrow HBrO_2 + Ce^{4+} \tag{R5}$$

Bromous acid breaks down to bromate ion and hypobromous acid,

$$2\,HBrO_2 \rightarrow BrO_3^- + HOBr + H^+ \tag{R6}$$

Subsequently hypobromous acid reacts with malonic acid to yield bromomalonic acid (= Bromination of malonic acid by hypobromous acid)

$$HOBr + CH_2(COOH)_2 \rightarrow BrCH(COOH)_2 + H_2O \tag{R8}$$

Process B is the total reaction as

$$2(R4) + 4(R5) + (R6) + (R8)\,, \quad \text{i.e.,}$$

$$\underset{\substack{\text{bromate}\\\text{ion}}}{BrO_3^-} + \underset{\text{cerium(III)}}{Ce^{3+}} + \underset{\substack{\text{malonic}\\\text{acid}}}{CH_2(COOH)_2} + 5\,H^+ \rightarrow \underset{\substack{\text{bromomalonic}\\\text{acid}}}{BrCH(COOH)_2} \\ + 4\,Ce^{4+} + 3\,H_2O \qquad (B) \\ \underset{\text{cerium(IV)}}{}$$

Furthermore, cerium(IV) produced in process (B) reacts with the organic species by the overall processes below:
Reaction (R9) (= Oxidation of Malonic Acid by Cerium[IV]):

$$6\,Ce^{4+} + CH_2(COOH)_2 + 2\,H_2O \rightarrow 6\,Ce^{3+} + HCOOH \\ + 2\,CO_2 + 6\,H^+ \tag{R9}$$

Reaction (R10) (= Oxidation of Bromomalonic acid by Cerium[IV]):

$$4\,Ce^{4+} + BrCH(COOH)_2 + 2\,H_2O \rightarrow Br^- + 4\,Ce^{3+} + HCOOH \\ + 2\,CO_2 + 5\,H^+ \tag{R10}$$

As the concentration of *bromomalonic acid*, $BrCH(COOH)_2$ increases reaction (R10) becomes important. The *bromide ion*, Br^- produced by (R10) is destroyed by (R2) as long as the autocatalytic production of bromous acid, $HBrO_2$, by the sequence (R4) + 2(R5) is able to maintain *bromous acid*. The concentration of $HBrO_2$ at the value of the second-order destruction of $HBrO_2$ by (R6), which creates the steady state is given by

$$(HBrO_2)_B = (k_4/2k_6)\,(BrO_3^-)\,(H^+) = 1 \times 10^{-4}\,(BrO_3^-)\,(H^+)$$

The concentration of bromous acid in Process B is over 10^5 times as great as that in Process A for which the bromous acid attains the steady state given by

$$(HBrO_2)_A = (k_1/k_2)\,(BrO_3^-)\,(H^+) = 5 \times 10^{-10}\,(BrO_3^-)\,(H^+)$$

When the rate of (R10) becomes sufficiently large, $HBrO_2$ drops rapidly to the value of given by $(HBrO)_A$, then Process B is turned off. As a result (Br^-) rises rapidly until the rate of (R10) is balanced by the rate of destruction by Process A initiated by (R1), and the oscillatory cycle starts again.

13

The overall *Process C* does not include the contribution of the small amout of concentration of cerium species compared to bromate, BrO_3^- and malonic acid $CH_2(COOH)_2$. Process C is the result of the sequence

$$x(A) + (3 - x)(B) + (2 - 2x)(R9) + 2x(R10)$$

where x may be between 0 and unity.

$$3 BrO_3^- + 5 CH_2(COOH)_2 + 3 H^+ \rightarrow 3 BrCH(COOH)_2$$
$$+ 2 HCOOH + 4 CO_2 + 5 H_2O \qquad (C)$$

bromate	malonic	formic
ion	acid	acid

It is shown in Field, Koros and Noyes (1972) that (R1) is related to the reaction of bromate with bromide in acid to give bromine

$$BrO_3^- + 5 Br^- + 6 H^+ \rightarrow 3 Br_2 + 3 H_2O$$

which was studied by Bray and Liebhafsky (1935), and A. Skrabal and S. R. Weberitsch (1915).

On the basis of these ten reactions a reaction mechanism of the oscillating reaction is described. All reactions are considered to be irreversible with the exception of reactions (R3), (R4) and (R5).

Reaction Scheme: Field and Noyes (1974)

Field and Noyes give the following simplified set of reactions of a model for reactions R_1-R_{10}, by taking the following combinations only:

$$A + Y \overset{k_1}{\rightleftarrows} X \qquad \text{corresponding to (R1)}$$
$$X + Y \overset{k_2}{\rightleftarrows} P \qquad \text{corresponding to (R2)}$$
$$B + X \overset{k_3}{\rightleftarrows} 2X + Z \qquad \text{corresponding to 2(R5) + (R4)}$$
$$2X \overset{k_4}{\rightleftarrows} Q \qquad \text{corresponding to (R4)}$$
$$Z \overset{k_5}{\rightleftarrows} fY \qquad \text{corresponding to (R10)}$$

where

$$X = HBrO_2, \text{ bromous acid,}$$
$$Y = Br^-, \text{ bromide,}$$
$$Z = Ce(IV), \text{ cerium(IV),}$$
$$A = B = BrO_3^- .$$

Differential Equations

If the steps of the above reaction scheme are assumed to be irreversible, then the following differential equations governing the reaction are obtained:

$$dx/dt = k_1 Ay - k_2 xy + k_3 Bx - 2k_4 x^2$$

$$dy/dt = -k_1 Ay - k_2 xy + f k_5 z$$

$$dz/dt = k_3 Bx - k_5 z \, .$$

Limit cycle oscillations obtained as a solution of this system are shown as projections of the x,y,z-phase space on the x,y and y,z phase planes, Fig. III.8a and III.8b, respectively.

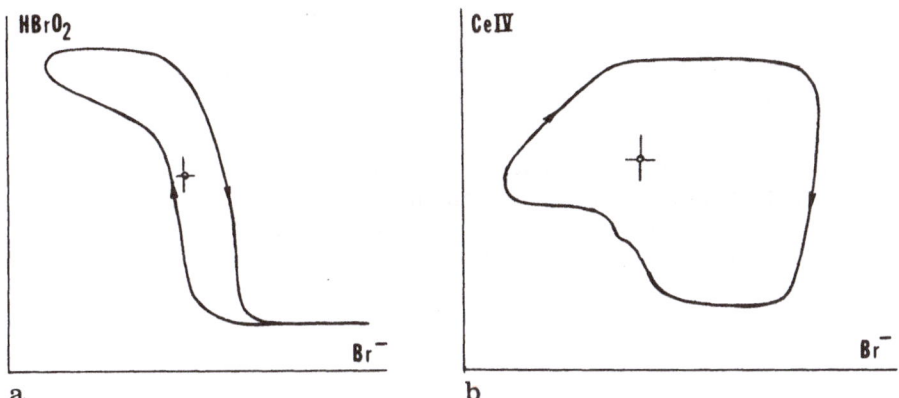

Fig. III.8a. Limit cycle behavior on Br⁻, HBrO₂-plane. (After Field and Noyes (1974)). **b** Br⁻, ce(IV) limit cycle. (After Field and Noyes (1974))

Bifurcation Analysis

The appearance and disappearance of the limit cycle behavior of this reaction are discussed in Field and Noyes (1974) in particular as related to the variables and parameters of the reaction system. Marek and Svobodova (1973) also discussed the appearance and disappearance of oscillatory behavior as well as the transition from one oscillatory solution to another. The onset of oscillations is further discussed in Berger and Koros (1980). These studies constitute in a way a nonsystematic analysis of the possible bifurcations of the system.

Mathematical Solutions

References: Zhabotinskii (1964-2), Vavilin, Zhabotinskii and Zaikin (1973), Zaikin and Zhabotinskii (1973), Marek and Svobodova (1975), Rossler and Wegmann (1978), Wegmann and Rossler (1978).

Mathematical models of the reaction yield various solutions. Some of the solutions obtained are: One singular point, 3 singular points, oscillating limit cycle, double periodic oscillations, chaotic oscillations.

Zhabotinskii and his colleagues, also Marek and Svobodova (1973) recognized the *double periodic* solutions, which were further applied by Boissonade in studying the Briggs-Rauscher oscillations, see Section III.B, Fig. III.9.

15

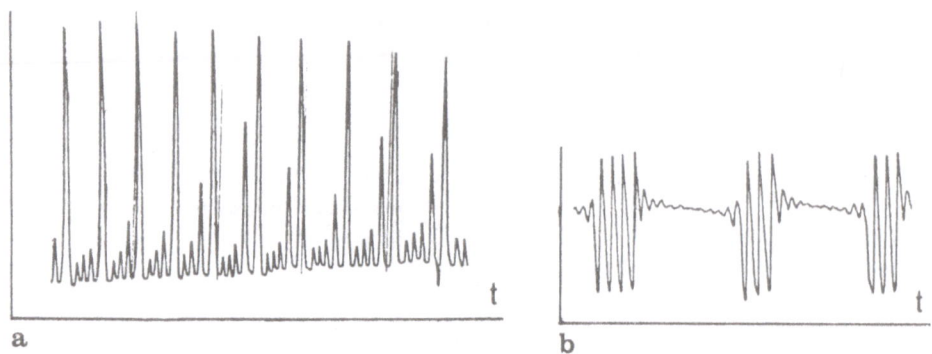

a

b

Fig. III.9. Double periodic oscillations, **a** (After Zhabotinskii (1964-2), **b** (After Marek and Svobodova (1975)

Zhabotinskii's observation of double oscillations was indeed the first indication of chaotic oscillations, see Figure 3 of Zhabotinskii (1964-2). Wegmann and Rossler (1978) observed the oscillations of electrochemical potential, a variable, in this reaction, Fig. III.10. Referring to the Model (1976-2) of Rossler, see page 44, they explained this chaotic behavior of the reaction as an example of the chaotic solution of the abstract model, Fig. III.10b.

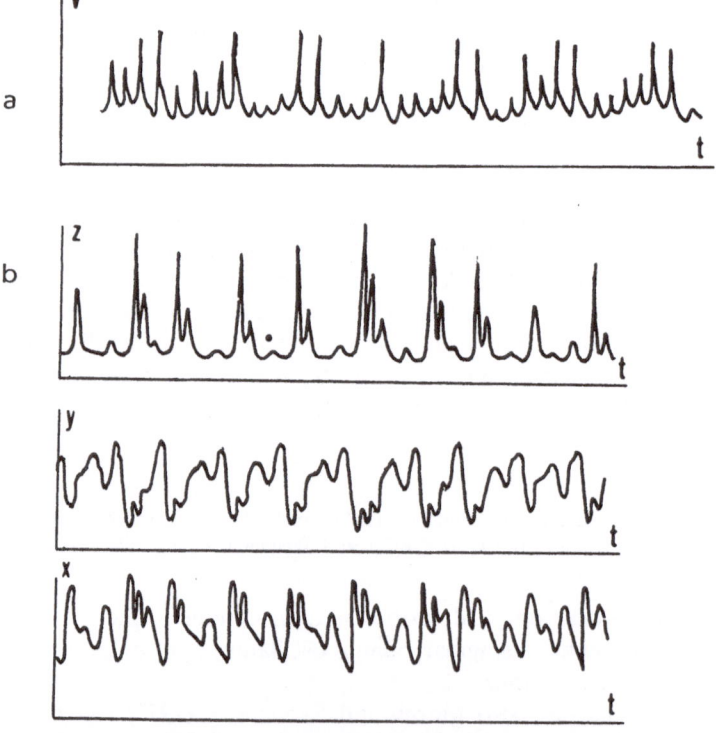

a

b

Fig. III.10. a Oscillations of electrochemical potential, **b** Oscillations of the model (1976-2). (After Wegmann and Rossler (1978))

Furthermore, Wegmann and Rossler experimenting with this reaction, observed oscillations corresponding to a) limit cycle, b) double limit cycle, and c) endogenous chaos and d) screw type chaos, see Fig. III.11, a–d, respectively.

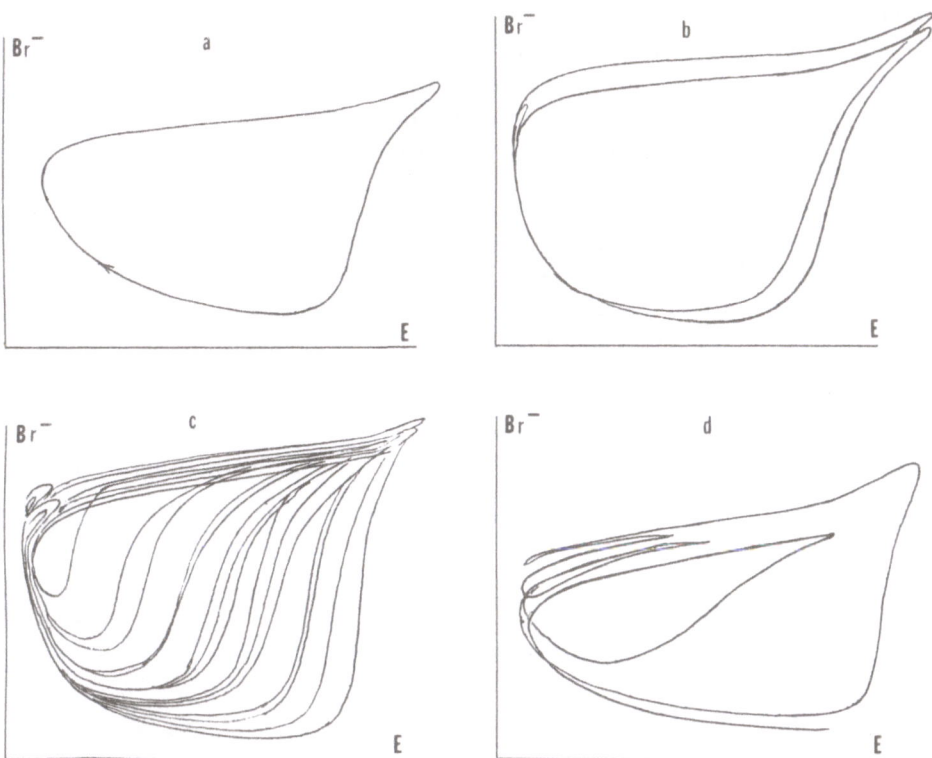

Fig. III.11. Experimental observations of **a** Limit cycle, **b** Double limit cycle, **c** Endogenous chaos, **d** Screw type chaos. Abscissa electrochemical potential, ordinate potential of bromide ion. (After Wegmann and Rossler (1978))

D Continuous Stirred Tank Reactor (CSTR)

References: Salnikov (1948), Bilous and Amundson (1955), Aris and Amundson (1958), Gurel and Lapidus (1965).

CSTR 1: Bilous and Amundson (1955)
Salnikov specifically reported multiple singular points and a limit cycle establishing the existence of oscillations in chemical reactions. Bilous and Amundson (1955) referred to Salnikov's (1948) paper as the first work where periodic phenomenon in reaction systems was discussed. They also indicated that a reaction A → B in CSTR is irreversible, exothermic, and kinetically first order. Considering mass balance and heat balance equations it is known that at the steady states, the heat consumption

17

equals the heat generation. Therefore from the mass and heat balance equations, one obtains

$$qcr(T - T_0) - U(T' - T) = -V(\Delta H)\, pe^{-E/RT} qA_0/(q + Vpe^{-E/RT})$$

where

q = volumetric flow rate of the influent
c = specific heat, $\quad r$ = density
A_0 = concentration of A in the influent
T = temperature, $\quad T_0$ = influent temperature
U = product of area and heat transfer coefficient
V = reactor volume
T' = average coolant temperature in reactor cooling coil
H = heat of reaction ($-\Delta H > 0$, exothermic)
p = frequency factor in reaction velocity constant
E = activation energy
k = $pe^{-E/RT} > 0$, reaction velocity constant.

Differential Equations: Bilous and Amundson (1955)
The two differential equations for concentration A and temperature T may be written as

$$dA/dt = a_1 - b_1 A - c_1 Ae^{-E/RT}$$
$$dT/dt = a_2 - b_2 T + c_2 Ae^{-E/RT}$$

where

$$a_i,\, b_i,\, c > 0.$$

Mathematical Solutions

Bilous and Amundson discuss the *multiple solutions* of the system. They consider the following possibilities in the solution plane are considered:
One singular point, two singular points, three singular points and their stability, as well as stable periodic solutions (sustained oscillations).
 Stability of singular points via Liapunov's second method is also discussed.

CSTR 2: Aris and Amundson (1958)

A slightly different variation of the above model is studied by Aris and Amundson. The differential equations are of the form:

$$dx/dt = 1 - x - P(x, y)$$
$$dy/dt = a_2 - b_2 y - c_2(y - d_2)\,(y - e_2) - P(x, y)$$

where

$$P(x, y) = x\, e^{(d_2 - f_2 y^{-1})}$$

Mathematical Solutions

Aris and Amundson discuss *multiple solutions,* including stable and unstable limit cycles of the system also. Their particular interest in this case is in the control of the system thus avoiding possible oscillations. In this paper they refer to the *bifurcation theory* of Poincare specifically, and show that parameter c_2 plays the role of a bifurcation parameter in the system.

CSTR 3: Gurel and Lapidus (1965)

A further variation of the above discussed CSTR models is studied by Gurel and Lapidus where the differential equations are given as

$$dx/dt = -x + b_1((1 - x)^{b_2} e^{(b_3 y/(y-1))} - 1)$$

$$dy/dt = -b_4 y - b_5((1 - x)^{b_3} e^{(b_3 y/(y-1))} - 1) - b_6 y(b_7 - y).$$

Mathematical Solutions

The limit cycle of this system is a flat one, and it is shown that the limit cycle is stable from the inside and the outside, Fig. III.12.

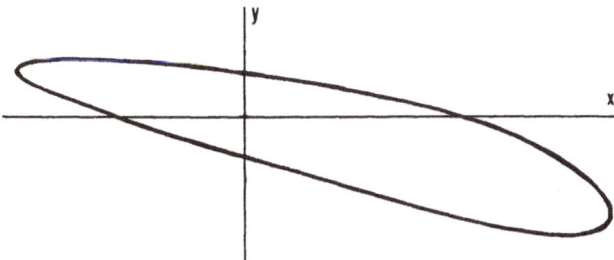

Fig. III.12. Limit cycle of the first order exothermic chemical reaction in CSTR (After Gurel and Lapidus (1965))

Solutions and Bifurcation Studies of CSTR

References: Uppal, et al. (1974), Uppal, et al. (1976), Ray (1977).

Later Uppal et al. (1974) applied the bifurcation theory to find all the solutions including the oscillatory solutions of the Continuous Stirred Tank Reactor. The CSTR equations that they studied are simplied form of those studied by Gurel and Lapidus (1965); more specifially the exponents b_2 and b_3 are equal to one:

$$dx_1/dt = -x_1 + Da(1 - x_1) \exp(x_2/(1 + (x_2/g)))$$

$$dx_2/dt = -x_2 + BD(1 - x_1) \exp(x_2/(1 + (x_2/g))) - b(x_2 - x_{2c}).$$

This bifurcation study has been reviewed at length by Ray (1977). The system possesses three singular points and limit cycles as oscillatory solutions. It should be noted that although the applications and the differential equations are significantly different from each other, some of the models of glycolysis studied by Sel'kov also possess three singular solutions and limit cycles very similar (topologically) to those obtained by Uppal, et al., see Section III.F.

Okan Gurel and Demet Gurel

E Solid Catalyzed Reactions

Catalytic Decomposition of N_2O

References: Hugo (1968), Hugo (1970).

The first experimental investigation in a solid catalyzed reaction system was reported by Hugo (1968). He observed periodic fluctuations in the exothermic decomposition of N_2O on a CuO catalyst.

Catalytic Oxidation of CO

References: Hugo (1970), Hugo and Jakubith (1972).

Experimenting with CO oxidation on a platinum screen (catalytic wire) under isothermic conditions Hugo (1972) reported observing undamped oscillations. Although these oscillations may be explained on the basis of a complicated reaction mechanism involving concentrations of the activated molecules, this is not elaborated in Hugo's paper.

References: Beusch et al. (1972).

Beusch et al. reported their experimental findings and exhibited the oscillations in CO_2 concentration as in Fig. III.13.

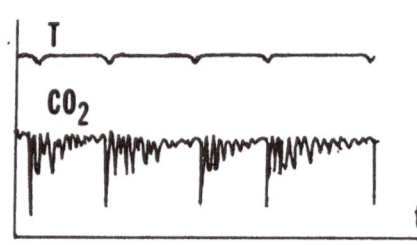

Fig. III.13. Oscillations in T, temperature and CO_2 concentrations (After Beusch (1972))

Reaction Mechanism:

References: Eigenberger (1976).

Based on the work of Beusch et al. Eigenberger (1976) repbrted a kinetic mechanism to explain the reaction as:

$$CO + s \rightarrow CO_s$$
$$O_2^{\cdot} + 2s \rightarrow O_{2s}$$
$$2 CO_2 + O_{2s} \rightarrow 2 CO_2 + 4s$$
$$P + s \rightarrow P_s$$

where P is an inhibitor. The mathematical model is constructed on kinetic rates only, and the physical transport process is not considered. The differential equations were

20

given for CO and P concentrations as variables, and a *limit cycle* in CO-P plane was obtained.

Reference: Eckert et al. (1973).

Carbon monoxide oxidation by pure oxygen on a porous catalyst of the type CuO on Al_2O_3 was studied in a laboratory differential recycle reactor. Under certain experimental conditions sustained oscillations of the catalyst bed temperature and CO concentrations in the reactor described in article I of Eckert et al. were observed and reported in article II of the series.

Balance equations in dimensionless form are

$$dx/dt = -x + Da(1 - x) E(T)$$
$$dT/dt = -mT + Da(1 - x) E(T)$$

where

$$E(T) = \exp\left(\frac{-T}{1 + T/g} + b\right)^{-1}$$

Da Damkohler number
T   dimensionless temperature
x   concentration of CO
m   parameter
g   dimensionless activation energy .

Experimentally observed observations are given in Fig. III.14.

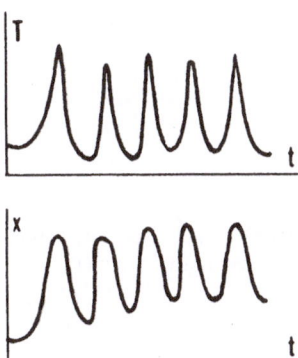

Fig. III.14. Experimentally observed oscillations in x and T. (After Eckert et al. (1973))

For two-component inlet mixture the equations become

$$dx/dt = -x + Da(1 - x)^{k2} x^{k3} E(T)$$
$$dT/dt = -mT + DaB(1 - x)^{k2} x^{k3} E(T) .$$

Okan Gurel and Demet Gurel

A Model for Oxidation of Carbon Monoxide (CO) in a Homogeneous System

Reference: Yang (1974).

Yang proposed a reaction scheme as:

$$CO + O + M \rightarrow CO_2^* + M \tag{1}$$
$$CO_2^* + O \rightarrow CO + O_2 \tag{2}$$
$$CO_2^* + M \rightarrow CO_2 + M \tag{3}$$
$$CO + OH \rightarrow CO_2 + H \tag{4}$$
$$H + O_2 \rightarrow OH + O \tag{5}$$
$$O + H_2O \rightarrow 2\,OH \tag{6}$$
$$H \rightarrow \text{destruction on wall} \tag{7}$$
$$O \rightarrow \text{destruction on wall} \tag{8}$$
$$OH \rightarrow \text{destruction on wall} \tag{9}$$
$$H + O_2 + M \rightarrow HO_2 + M \tag{10}$$

where $x = O$, $y = CO_2^*$, $z = OH$, $w = H$ concentrations. Differential equations obtained are

$$dx/dt = k_5 w - k_1 x - k_2 xy - k_6 x - k_8 x$$

$$dy/dt = k_1 x - k_2 xy - k_3 y$$

$$dz/dt = \text{linear function of } w, z \text{ and } x \text{ only}$$

$$dw/dt = \text{linear function of } z \text{ and } w \text{ only.}$$

A simulation result exhibiting *limit cycle* oscillations is given in Fig. III.15.

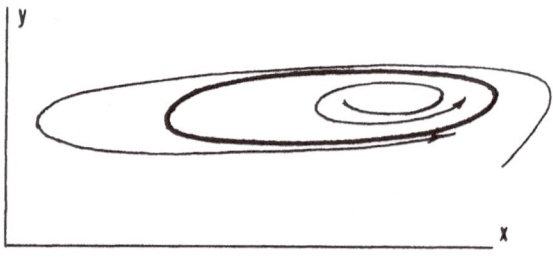

Fig. III.15. Limit cycle oscillations on x,y-plane. (After Yang (1974))

Catalytic Oxidation of Hydrogen

References: Belyaev et al. (1973), Horak and Jiracek (1972).

Belyaev, et al. studied the catalytic oxidation of hydrogen, and found that the reactants act on the catalyst. This effect may be regarded as *feedback*. Several stationary states were found for the *catalytic hydrogen oxidation* reactions on nickel and for the reaction of CO with H. Belyaev et al. found stable auto-oscillations of the hydrogen oxidation rate on nickel foil under isothermal conditions. See Fig. III.16.

Horak and Jiracek (1972) observed three steady states and indicated that the stability of steady states depends on the ratio of the reactor volume to the amount of catalyst.

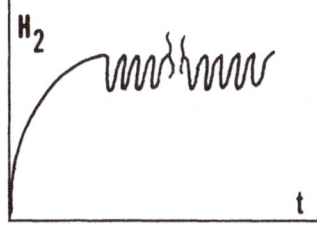

Fig. III.16. Experimental observation of oscillations in H_2. (After Balyaev (1973))

Reaction Mechanism

Reference: Pikios and Luss (1977).

Based on the following chemical scheme which they proposed, Pikios and Luss (1977) gave the balance equations.

$$A_1 + s \rightarrow A_{1s}$$
$$A_2 + s \rightarrow A_{2s}$$
$$A_{1s} = A_{2s} \rightarrow A_3 + 2s \,.$$

In their analysis they observed *limit cycle* oscillations.

Studying the catalytic oxidation of hydrogen, Horak and Jiracek (1972) reported the catalytic exothermic reaction between hydrogen and oxygen. A single pellet reactor with an external reservoire of a variable volume is used in the recirculation loop. For a fixed amount of catalyst in the reactor, the reaction between the heat and mass capacity can be varied by varying the volume of the extra reservoire. Varying the reservoire volume, Horak and Jiracek observed a *limit cycle* behavior for this reaction, Fig. III.17.

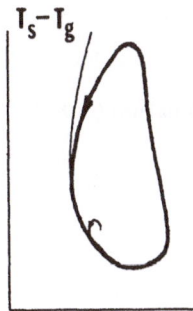

Fig. III.17. Limit cycle oscillations resulting from the effect of the reactor volume V on the stability of steady states. (After Hlavacek and Votruba (1978))

F Oscillations in Glycolysis

References: Duysens and Amesz (1957), Ghosh and Chance (1964), Chance, Hess and Betz (1964), Higgins (1964).

Oscillations in glycolytic system have been observed experimentally by numerous investigators. An early observation is by Duysens and Amesz (1957). By adding glucose (GLU), they observed oscillations in the concentration of reduced phosphopyridine nucleotide (NADH). Later, Chance et al. (1964) also reported NADH oscillations in a cell-free extract, and Ghosh and Chance observed oscillations in fructose-1,6-diphosphate (FDP) and glucose-6-phosphate (G6P), see Fig. III.18.

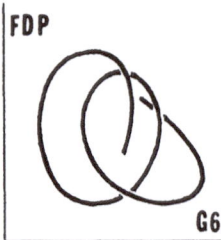

Fig. III.18. Oscillations in concentrations of FDP and G6P (From Ghosh and Chance (1964))

Reaction Mechanism

Reference: Higgins (1964).

Higgins formulated a generalized chemical mechanism for oscillating reactions. This mechanism was then used to explain glycolytic oscillations. Based on the known chemistry of phosphofructokinase (PFK) and the associated glycolytic intermediates, Higgins proposed:

$$GLU \rightarrow F6P$$
$$F6P + E_1 \rightarrow E_1 \cdot F6P$$
$$E_1 \cdot F6P \rightarrow E_1 + FDP$$
$$FDP + E_1^+ \rightarrow E_1$$
$$FDP + E_2 \rightarrow E_2 \cdot FDP$$
$$E_2 \cdot FDP \rightarrow E_2 + GAP$$

where

E_1 = phosphofructokinase (enzyme) (PFK)
$F6P$ = fructose-6-phosphate (substrate)
FDP = fructose-1,6-diphosphate (product)
E_2 = Aldolase and triose phosphate isomerase combination (enzyme)
GAP = glyceraldehyde phosphate (product) .

The above mechanism is summarized as

$$GLU \longrightarrow F6P \xrightarrow{} FDP \longrightarrow GAP$$
$$\text{substrate}$$

The basic oscillatory couple exhibiting a limit cycle is F6P-FDP. Moreover, oscillations in GLU are also observed. The end product is glyceraldehyde phosphate (GAP). The first step from GLU to F6P is a first-order reaction and in the second step, an activated form of phosphofructokinase acts as an enzyme. FDP activates this second step.

Models for Oscillating Reactions in Glycolysis

Model 1: Back Activation Model

Reference: Higgins (1967).

Higgins (1967) proposed various models for glycolysis, based on *feedback* mechanisms, such as back activation, back inhibition, forward activation and forward inhibition. Here we refer to the model based on *back activation* feedback. The reaction is represented by the equations:

$$dx/dt = k - \frac{axy}{y(b + x) + c} \, .$$

$$dy/dt = \frac{axy}{y(b + x) + c} - \frac{dy}{e + y}$$

where $x = $ F6P and $y = $ FDP.

This model possesses a stable singular point which bifurcates to yield a *stable limit cycle*.

Model 2:

Reference: Sel'kov (1968-1).

Sel'kov (1968) proposed the following kinetic model for the phosphofructokinase reaction in glycolysis.

$$\xrightarrow{v_1} S_1 + ES_2^g \underset{k_{-1}}{\overset{k_{+1}}{\rightleftarrows}} S_1 ES_2^g$$

$$S_1 ES_2^g \xrightarrow{k_{+2}} ES_2^g + S_2 \xrightarrow{v_2}$$

$$gS_2 + E \underset{k_{-3}}{\overset{k_{+3}}{\rightleftarrows}} ES_2^g$$

where

$S_1 = $ ATP, supplied at the rate v_1, irreversibly converted to S_2.
$S_2 = $ ADP, as product, which is removed at the rate V_2.
$E = $ PFK, (inactive) free enzyme.
  E becomes active by combining with g (ADP) product molecules to form the complex ES_2^g.
$g = $ The order of activation of PFK by glycolytic intermediates ADP and AMP (Adenosine mono phosphate).

25

Denoting $x_1 = ES^g$, $x_2 = S_1 ES_2^g$, $e = E$, $s_1 = S_1$, $s_2 = S_2$, the differential equations become:

$$ds_1/dt = v_1 - k_1 s_1 x_1 + k_{-1} x_2$$
$$ds_2/dt = k_2 x_2 - k_3 s_2^g e + k_{-3} x_1 - k_2 s_2$$
$$dx_1/dt = -k_1 s_1 x_1 + (k_{-1} + k_2) x_2 + k_3 s_2^g e - k_{-3} x_1$$
$$dx_2/dt = k_1 s_1 x_1 - (k_{-1} + k_2) x_2$$
$$de/dt = -k_3 s_2^g e + k_{-1} x_1 .$$

Under the assumptions of limit transition the last three derivatives are small, thus the model is reduced from five to two equations. Substituting,

$$x = s_1 = k_1 s_1/(k_{-1} + k_2) = ATP$$
$$y = s_2 = (k_3/k_{-3})^{1/g} s_2 = ADP ,$$

the differential equations become

$$dx/dt = v_1 - \frac{xy^g}{1 + y^g(1 + x)}$$

$$dy/dt = a\left(\frac{xy^g}{1 + y^g(1 + x)} - by\right)$$

At a verly low rate of the glycolytic flux, $v_1 \ll 1$, these equations are simplified as

$$dx/dt = 1 - xy^g$$
$$dy/dt = ay(xy^{g-1} - 1) .$$

For certain values of the parameters this system exhibits a stable limit cycle, Fig. III.19.

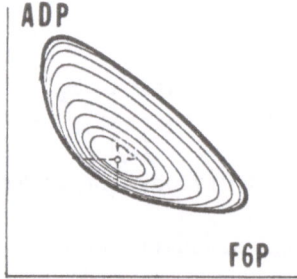

ADP

F6P

Fig. III.19. Limit cycle of Sel'kov model (1968) (From Sel'kov (1968-1))

Model 3:

References: Sel'kov (1968-2), Sel'kov and Betz (1973).

Revising (1968-1) model Sel'kov (1968-2) contructed a new model:

$$dx/dt = n - B$$

$$dy/dt = b(B - cy)$$

where $x =$ F6P, $y =$ ADP and

$$B = \frac{x(m + y^g)}{1 + ax + y^g(1 + x)}$$

This model was slightly altered by Sel'kov and Betz (1973). The model still has only one singular solution, and a stable limit cycle corresponding to oscillations.

Although the earlier models by Sel'kov exhibit only one limit cycle, later models have multiple solutions, both oscillatory and nonoscillatory. These models are quite interesting for the reason that these are the earliest models in biochemical oscillations giving clear evidence for more complex solutions than a simple limit cycle as has been the case with almost all other chemical systems except those in CSTR models.

Model 4: Multiple Singular Points and Multiple Limit Cycles

References: Sel'kov (1972), Kaimachnikov and Sel'kov (1975), Sel'kov, Dynnik and Kirsta (1979).

a) Another early model, Sel'kov (1972), is based on the reaction $X + Y' \rightarrow X' + Y$ catalyzed by the enzyme E, inhibited by the active form of the coenzyme Y'.

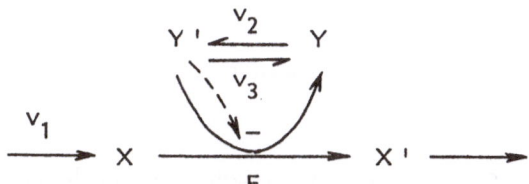

The differential equations are

$$dx/dt = v_1 - v$$

$$m \, dy/dt = d - y' - v$$

where

$$v = \frac{xy'}{z + y'(1 + ay'^g)}$$

$$v_1 = v_m - bx$$

This model yields various solutions. Depending on variations in parameters, the following cases are obtained:

Case 1. One singular point
Case 2. Three singular points surrounded by a stable limit cycle
Case 3. Solutions in Case 2, with additional unstable limit cycles.

Further variations of this model are also presented in the same paper.

b) *Open Enzymatic Reaction Scheme*: Kaimachnikov and Sel'kov considered the following reaction scheme which is a variation of Sel'kov (1972) model.

$$S + E \underset{k_{-1}}{\overset{k_{+1}}{\rightleftarrows}} SE$$

$$SE + A \underset{k_{-3}}{\overset{k_{+3}}{\rightleftarrows}} SE_a \overset{k_{+2}}{\rightarrow} P + E + B$$

$$SE + gS \underset{k_{-4}}{\overset{k_{+4}}{\rightleftarrows}} SES^g$$

$$SE_a + gS \underset{k_{-4}}{\overset{k_{+4}}{\rightleftarrows}} SE_aS^g$$

where

S = substrate which inhibits E
E = enzyme
A, B = forms of coenzymes
P = product

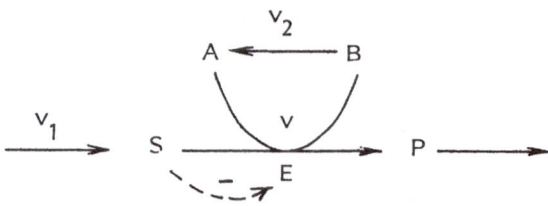

v_1 is the rate of formation of S, v_2 is the rate of conversion of the inactive form B to the reacting form. v is the reaction rate of $S + A \rightarrow P + B$.

Substituting x = S and y = A the differential equations become:

$$dx/dt = v_0 - b_1 x - v$$

$$dy/dt = 1 - y - v$$

where

$$v = \frac{axy}{(1 + dy)(c + x(1 + x^\theta))}$$

Solutions obtained are multiple limit cycles, and multiple singular points as shown in Fig. III.20.

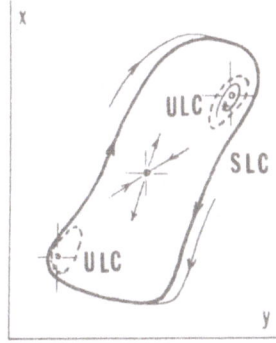

Fig. III.20. Multiple limit cycles, one stable and two unstable, and multiple singular points, two stable focus and one saddle point. (From Kaimachnikov and Sel'kov (1975))

c) Another model with multiple solutions was also proposed by Sel'kov and his colaborators, Sel'kov et al. (1979). Here the reaction scheme is given as

$$ATP + Fructose\text{-}6P \rightarrow Fructose\text{-}1,6\text{-}P_2\text{-}ADP$$

$$H_2O + Fructose\text{-}1,6\text{-}P_2 \rightarrow Fructose\text{-}6\text{-}P + P_{inorg}$$
$$Fructose$$
$$biphosphatase$$
$$(FBPase)$$

The differential equations are:

$$dx/dt = v_{1m} - k_1 x - v$$
$$e\, dy/dt = v_{2m} - k_2 y + v$$

where $x = $ F6P, and $y = $ FBP, and

$$v = \frac{x}{1 + sx} - \frac{y}{c + y(1 + y_d)}$$

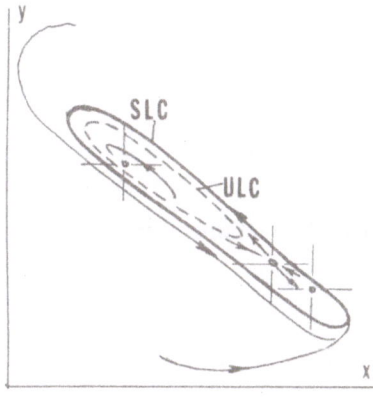

Fig. III.21. One unstable limit cycle around a stable singular point and one stable limit cycle around the singular point and the unstable limit cycle. (From Sel'kov, Dynnik and Kirsta (1979))

29

Okan Gurel and Demet Gurel

In this model it is shown that a stable singular point inside a stable limit cycle is separated by an unstable limit cycle, see Fig. III.21.

Model 5: Coexistence of a Stable Limit Cycle and a Stable Singular Point

Reference: Dynnik, Sel'kov and Semashko (1973).

In this model, the glycolytic step involving reactions catalyzed by phosphofructo-kinase (PFK), adenylatekinase (ADK) and a generalized enzyme, E_e, is considered.

Reactions Scheme

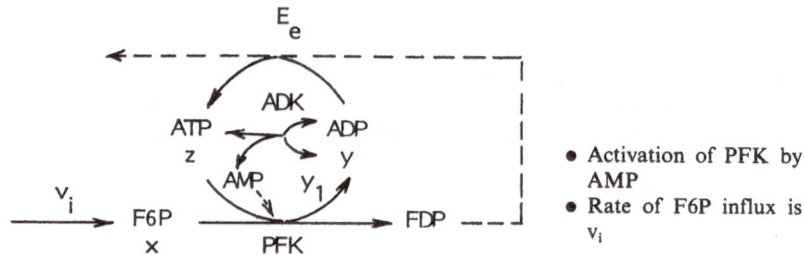

● Activation of PFK by AMP
● Rate of F6P influx is v_i

Various reactions are,

1. Phosphorylation of fructose-6-phosphate (F6P) catalyzed by phosphofructokinase (PFK)
2. Reversible phosphorylation of ADP catalyzed by adanylatekinase (ADK)
3. Phosphorylation of ADP catalyzed by the enzyme E_e whose action is equivalent to a total effect of the enzymes of the glycolytic steps and oxidative phosphory-lation.

The differential equation are:

$$dx/dt = v_i - v$$
$$dy_1/dt = v_a/a$$
$$dy_2/dt = -2v_a/a + v - y_2$$
$$dz/dt = v_a/a - v + y_2$$

where

$$v = x \frac{z}{b+z} \cdot \frac{c+y_1}{d+y_1}$$

By changing variables as $y = y_1 + y_2/2$ the differential equations become

$$dx/dt = v_i - v$$
$$dy/dt = v - y_2$$
$$a \, dy_1/dt = v_1$$

30

and as $a \to 0$,

$$dx/dt = v_i - v^*(x, y)$$
$$dy/dt = v^*(x, y) - y_2$$

where v^* is calculated from the expression for v above. There are two stationary states S_1 and S_2, one a stable (or unstable) focus, the other a saddle point.

The solutions obtained are:

1. No singular solution in the finite domain.
2. One singular solution.
3. Two singular solutions, one saddle (S_2), one unstable focus (S_1) and a stable limit cycle.
4. S_1 stable, S_2 saddle, and S_1 surrounded by an unstable limit cycle.
5. S_1 stable, S_2 saddle, S_1 surrounded by unsable limit cycle which is surrounded by a stable limit cycle.
6. S_1 unstable, S_2 saddle.

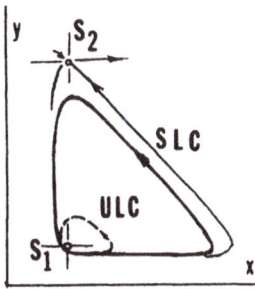

Fig. III.22. Coexistence of stable limit cycle and stable singular point surrounded by an unstable limit cycle, (After Dynnik et al. (1973))

Model 6:

Reference: Dynnik and Sel'kov (1975-1).

Kinetic Model

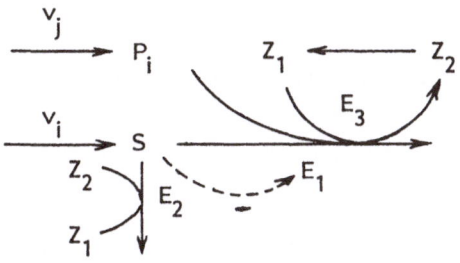

S   = glyceraldehyde phosphate (GAP).
P_i = inorganic phosphate.
Z_1, Z_2 = coenzymes NAD^+ and NADH.

31

v_j = Constant rate of injection of P_i.
v_i = V_m — K_0S rate of injection of S.
V_m and K_0 = maximum rate of injection and the rate of constant of leakage of S.

Reactions catalyzed by aldolase (ALD) and triosophosphate isomerase (TPI) are considered rapid, and in equilibrium. Therefore, fructose-1,6-diphosphate (FDP), glyceraldehyde phosphate (GAP) and dihydroxyacetone phosphate (DHAP) change in phase, thus may be described by

$$DHAP = R_1 (GAP)$$
$$FDP = R_1 R_2 (GAP)^2$$

where R_1 and R_2 are equilibrium constants of the corresponding reactions.
The differential equations are:

$$dP/dt = v_j - v_2$$
$$e \, dG/dt = v - v_1 - v_2$$

where

$$P = P_i, \qquad G = S = (GAP)$$
$$v_1 = PG/(1 + G + aG^g)$$
$$v_2 = bG \, .$$

The limit cycle obtained as a solution is shown in Fig. III.23.

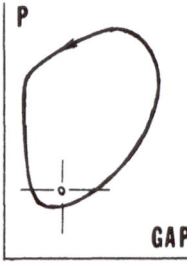

Fig. III.23. The simple limit cycle (From Dynnik and Sel'kov (1975-1))

Model 7: Double Periodic Limit Cycle

Reference: Pye and Chance (1966), Dynnik and Sel'kov (1975-2).

Oscillations in NADH concentrations as well as in other glycolytic intermediates are observed experimentally by Pye and Chance (1966). In addition double-frequency oscillations in NADH are observed, Fig. III.24.

NADH

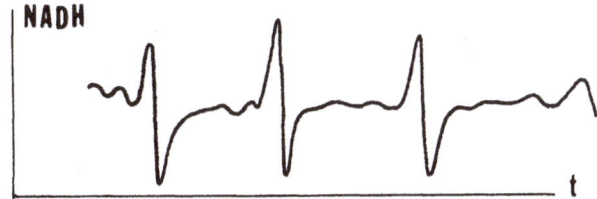

Fig. III.24. Double oscillations in NADH (From Pye and Chance (1966))

Kinetic Model

Dynnik and Sel'kov proposed the following reaction scheme which leads to double periodic solutions:

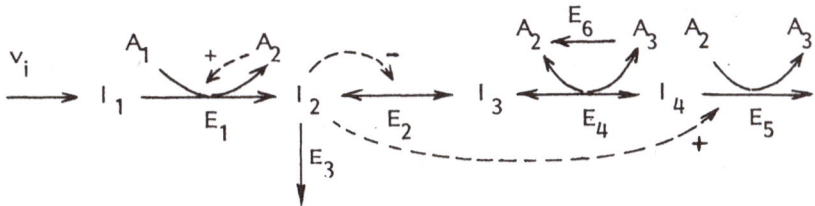

where

E_1 = PFK
E_2 = glyceraldehyde phosphate dehydrogenase (GAPDH)
E_3 = α-glycerophosphate dehydrogenase (αGPDH)
E_4 = phosphoglycerate kinase (PGK)
E_5 = pyruvate kinase (PVK)
E_6 = generalized ATPase
I_1 = fructose-6-phosphate (F6P)
I_2 = glyceraldehyde phosphate (GAP),
 dihydroxyacetone (DHAP),
 fructose-1,6-diphosphate (FDP) mixture
I_3 = 1,3-diphosphoglycerate (1,3 DPG)
I_4 = phosphoenolpuryvate (PEP),
 2-phosphoglycerate (2PG),
 3-phosphoglycerate (3PG) mixture
A_2, A_3 = ADP, ATP .

Dotted lined arrows indicate activation of PFK by the product of ADP, activation of PVK by FDP and inhibition of GAPDH by the substrate GAP.

The Mathematical Model

The reaction rates are given as follows:

$$v_1 = X(S^{g_1} + v_0)/(S^{g_1} + 1)$$
Rate of PFK reaction (Sel'kov and Betz (1973))

33

$$v_2 = G/(1 + G + a_1 G^{g_2}) - b_1 Z$$
$\qquad$ Rate of GAPDH reaction (Dynnik and Sel'kov (1975-1))

$$v_3 = aG$$
$\qquad$ Rate of α-GPDH (Dynnik and Sel'kov (1975-1))

Rates of reactions catalyzed by E_4, E_5 and E_6 are

$$v_4 = SZ/(S + n) - b_2 Y$$

$$v_5 = b_3 YS(G^{g_3} + w_0)$$

$$v_6 = d = \text{constant} .$$

Excluding the rapid varying Z, the model is reduced to:

$$e \ dX/dt = v_i - v_1$$

$$e_1 \ dS/dt = rv_1 - v_2^* - v_5 + v_6$$

$$e_2 \ dG/dt = 2rv_1 - v_2^* - v_3$$

$$e_4 \ dY/dt = v_2^* - v_5$$

where

$$v_2^* = \frac{G}{1 + G + a_1 G^{g_2}} \frac{f}{b_1} - b_2 Y$$

$X = \text{F6P} (= I_1)$

$S = \text{ADP} (= A_2)$

$G = \text{GAP, DGAP, FDP mixture} (= I_2)$

$Y = \text{PEP, 2PG and 3PG mixture (PEV)} (= I_4) .$

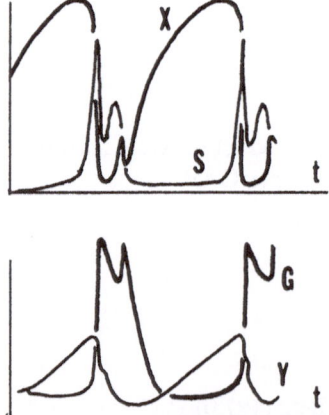

Fig. III.25. Double oscillations in F6P, ADP, GAP and PEV (From Dynnik and Sel'kov (1975-2))

Resulting solutions exhibit double oscillations, Fig. III.25.

Model 8: Complex Behavior in Glycolysis

References: Schulmeister and Sel'kov (1978), Schulmeister (1978), Kaimachnikov and Schulmeister (1979), Sel'kov (1980).

a) Schulmeister and Sel'kov considered an open self oscillating enzyme reaction as $X + Y \rightarrow X' + Y'$, coupled with a reversible enzymatic "deposition" of X.

The differential equations are: The mathematical model proposed consists of three differential equations:

$$dx/dt = v_1 - v - dv_a$$

$$n\, dy/dt = v_2 - v$$

$$dz/dt = m\, dv_a$$

where

$$v_1 = v_{1m} - bx\,, \qquad v_2 = v_{2m} - hy\,,$$

$$v = xy(e + y(1 + ay^c))^{-1}$$

$$v_a = x(g + x)^{-1} - fz(1 + z)^{-1}\,.$$

Mathematical Solutions

An interesting solution obtained is a *folded limit cycle*, which is a periodic solution with more than one maximum, see Fig. III.26.

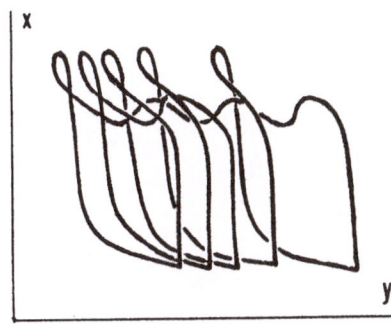

X

y

Fig. III.26. Folded limit cycle. (From Schulmeister and Sel'kov (1978))

b) Another model by Schulmeister (1978) is proposed as

$$dx/dt = v_1 - v - v_d$$

$$dy/dt = a(v + v_2)$$

$$dz/dt = fv_d$$

35

where

$$v_1 = 1 - bx, \qquad v = xy^2, \qquad v_2 = d - y, \qquad v_d = exy - z$$

x = glucose-6-phosphate, y = fructose-1,6-diphosphate, z = glycogen. In this model a chaotic solution is obtained, Fig. III.27.

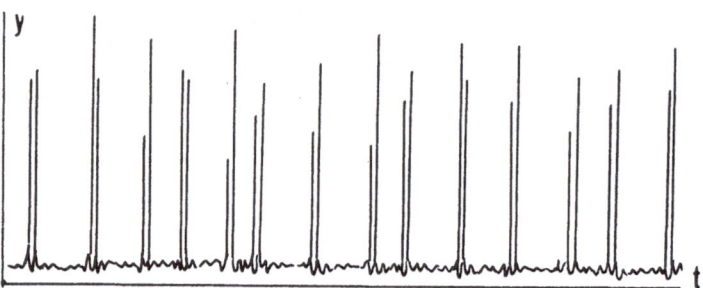

Fig. III.27. A chaotic solution in glycolysis (From Schulmeister (1978))

c) Another complex behavior observed by Sel'kov (1980), is given in Fig. III.28.

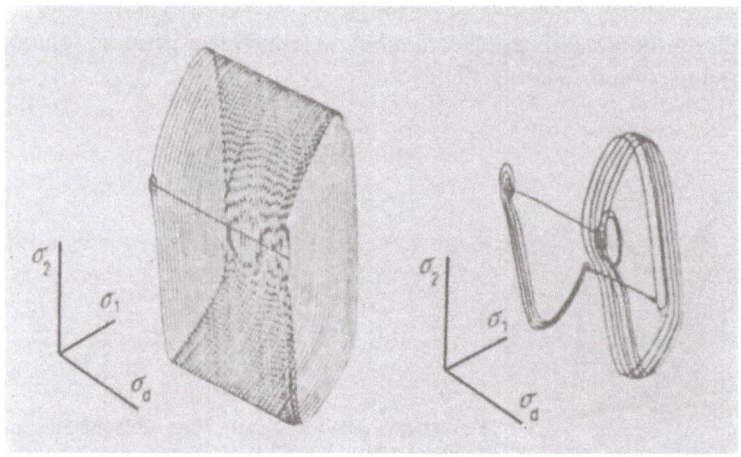

Fig. III.28. Complex behavior in glycolysis, (From Sel'kov (1980))

d) An open mono substrate enzymatic reaction with substrate inhibition and reversible deposition of product has been examined and reported by Kaimachnikov and Schulmeister (1979):

Reaction Scheme

$$\xrightarrow{\ v_1\ } \text{S} \xrightleftharpoons{\ v\ } \text{P} \xrightarrow{\ v_2\ }$$

$$v_{-d} \Big\Uparrow\Big\Downarrow v_{+d}$$

$$\text{P}_d$$

where S = substrate, P = product, P_d = deposed form of the product, v_1 and v_2 = influx rate of S and outflux rate of P, v = reaction rate, v_{+d}, v_{-d} = forward and backward reaction rates for the deposed form, respectively.

Mathematical Model

$$dx/dt = v_{1m} - b_1 x - v$$

$$dy/dt = v_{2m} - b_2 y + v - m(y - z)$$

$$dz/dt = ma(y - z)$$

where

$$v = \frac{x - ky}{1 + (x + y)(1 + cx^g)}$$

x = S, y = P, z = P_d, v_m = maximum reaction rate.

The phase trajectories for $b_1 = 0.026$, $v_{1m} = 0.0659$ are given in Fig. III.29 for m values as A) 0.023, B) 0.0234, C and D) 0.02365, E) 0.025 and F) 0.0254.

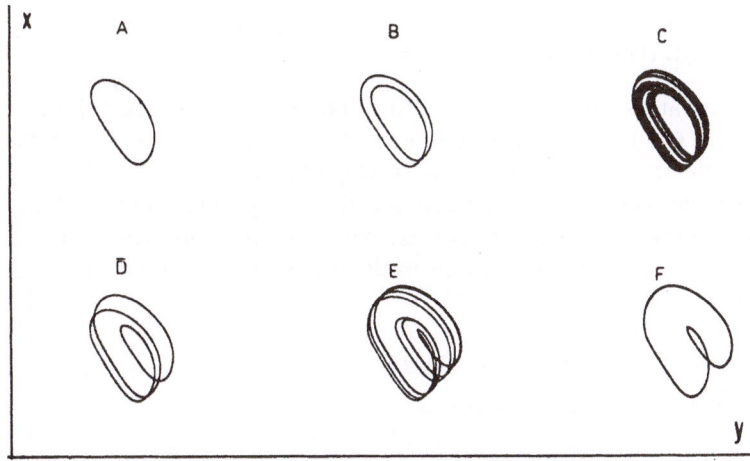

Fig. III.29. (From Kaimachnikov and Schulmeister (1979))

When the results in Fig. III.29 B, D, E, F are compared with the generic multi periodic limit cycle in Fig. IV.4 the relationship between the model solutions and the generic possibilities can be easily seen.

Another set of solutions by Kaimachnikov and Schulmeister exhibiting complex behavior are given in Fig. III.30.

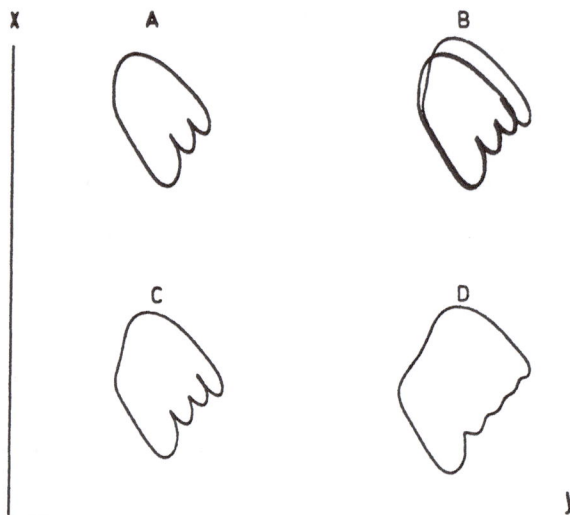

Fig. III.30. (From Kaimachnikov and Schulmeister (1979))

G Peroxidase Catalyzed Reactions

Horseradish Peroxidase Catalyzed Oxidation of Reduced Pyridine Nucleotides (NADH)

References: Yamazaki, et al. (1965), Yamazaki and Yokota (1967), Degn (1968), Degn (1969), Degn and Mayer (1969), Olsen and Degn (1977).

Yamazaki and collaborators studied the horseradish peroxidase catalyzed oxidation of NADH by oxygen, O_2, in an open system, and observed oscillations in concentrations of oxygen (1965) and of NADH (1967). Degn (1968, 1969) also observed oscillations in a similar peroxidase catalyzed reaction. Degn and Mayer (1969) investigated the theoretical considerations of these reactions and more recently Olsen and Degn (1977) observed chaotic oscillations in the peroxidase catalyzed oxidation of NADH by O_2 in an open system, Fig. III.31.

Lactoperoxidase Catalyzed Oxidation of NADPH

Reference: Nakamura, Yokota and Yamazaki (1969).

Nakamura et al. (1969) reported the oscillatory oxidation of NADPH in a lactoperoxidase system, and observed sustained oscillations in oxygen concentration.

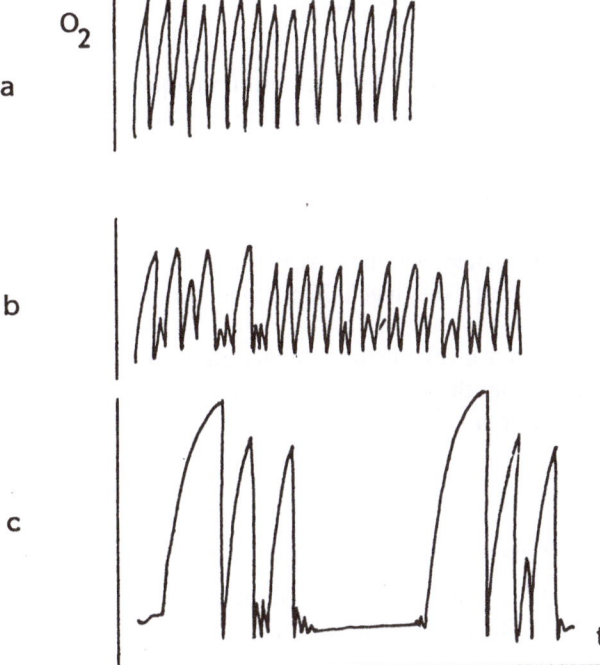

a

b

c

Fig. III.31. Chaotic oscillations in horesradish peroxidase catalyzed oxidation of NADH by O_2. Cases a), b) and c) are for decreasing concentrations of peroxidase. (From Olsen and Degn (1977))

Thyroid Peroxidase Catalyzed Iodination of Thyroglobulin

References: D. Gurel (1975), D. Gurel and Gans (1976), D. Gurel and McNelis (1977).

It was observed by D. Gurel (1975) that enzymatic iodination of thyroglobulin and coupling of iodinated tyrosyl residues to yield protein bound thyroid hormone have the elements of the Bray reaction and are similar to enzyme catalyzed reactions exhibiting oscillatory behavior. Based on Sel'kov (1972) model, the following reaction scheme for the iodination of thyroglobulin reaction was proposed by D. Gurel (1975), Gurel and Gans (1976), Gurel and McNelis (1977):

where TPO = Thyroid peroxidase, TYR = Tyrosyl residues on thyroglobulin, TYR-I = Iodinated TYR. The dotted arrow indicates the inhibition of iodination reaction by iodide, and represents the Wolff-Chaikoff effect.

39

H Decomposition of Sodium Dithionite (Na₂S₂O₄)

References: Lotka (1910-1), Lotka (1910-2), Rinker, et al. (1965), DePoy and Mason (1974).

Sodium dithionite ($Na_2S_2O_4$) undergoes thermal decomposition in aqueous solution to form sodium bisulphite ($NaHSO_3$), and sodium thiosulphate ($Na_2S_2O_3$) according to the stoichiometry:

$$2\,Na_2S_2O_4 + H_2O \rightarrow 2\,NaHSO_3 + Na_2S_2O_3$$

Rinker et al. reported that at certain temperatures oscillations in the dithionite concentrations are observable. However, the mechanism they proposed did not account for the oscillations.

DePoy and Mason (1974) proposed a mechanism including an autocatalytic step which produces oscillations, and gave a mathematical model with product nonlinearity as in the case of the Lotka model, Lotka (1910-1, 2).

I Bimolecular Model

Reference: Lefever (1968), Lefever and Nicolis (1971).

A hypothetical bimolecular model was proposed by Lefever (1968). For certain values of parameters this model exhibits an oscillatory solution. This oscillatory solution is represented by a *limit cycle*.

Reaction scheme: Lefever (1968)

This reaction scheme consists of a set of reactions between the two intermediate substances X and Y as follows:

$$A \rightleftarrows X$$
$$B + X \rightleftarrows Y + D$$
$$2X + Y \rightleftarrows 3X$$
$$X \rightleftarrows E$$

Differential Equations

Rate equations lead to two nonlinear differential equations for X and Y concentrations. The nonlinearity of the system is due to simple power terms such as x^2y.

$$dx/dt = k_1 A - k_2 Bx + k_3 x^2 y - k_4 x$$
$$dy/dt = k_2 Bx - k_3 x^2 y\,.$$

Solutions

Reference: Lefever and Nicolis (1971).

This model possesses one stable singular point which subsequently bifurcates into an unstable singular point and a surrounding stable limit cycle, Figure III.32.

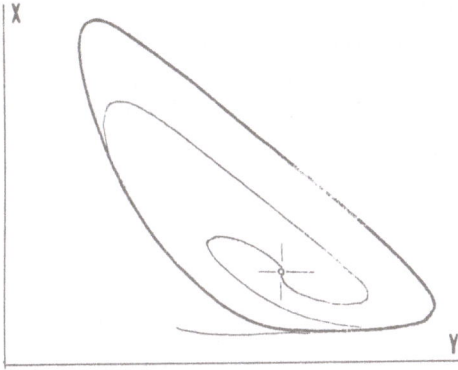

Fig. III.32. Limit cycle of the Brussels model. (After Lefever and Nicolis (1971))

Bifurcation Analysis

The equations are quite simple and the bifurcation analysis can be carried out in a straight forward fashion. The only singular point of the system is at $A + X^2 Y - X - BX = 0$, $BX - X^2 Y = 0$, leading to $X_0 A$ and $Y_0 = B/A$.

The coefficient matrix of the linearized equations at the X_0, Y_0 is:

$$\begin{vmatrix} -1 + B & A^2 \\ -B & -A^2 \end{vmatrix}$$

The three Poincare conditions, see Gurel (1979-2) are:
1. The determinant of the coefficient matrix vanishes, $A^2 = 0$, thus this condition is satisfied for $A = 0$.
2. The stability change is satisfied by the characteristic equation,

$$((-1 + B) - s)(-A^2 - s) + BA^2 = 0$$

which results in

$$2s_{1,2} = B - 1 - A^2 + (or\ -)\ ((B - 1 - A^2)^2 - 4A^2)^{1/2}.$$

$s = 0$ implies $B = 1 + A^2$, thus at the parameter values $A = 0$, $B = 1$.

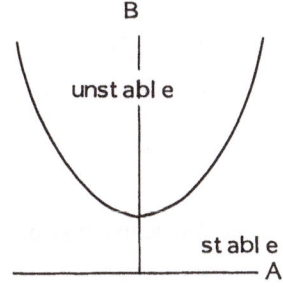

Okan Gurel and Demet Gurel

3. There are in fact two solutions after the singular solution, $X = A$, $Y = B/A$ bifurcates, the new solution is a stable limit cycle. Lefever and Nicolis (1971) give these solutions for the point with parameter values $A = 1$, $B = 3$, lying in the parameter plane region corresponding to a limit cycle and an unstable singular point, simultaneously.

J Abstract Reaction Systems and Their Models

References: Rossler (1975 through 1980).

In a series of articles, Rossler proposed *abstract models* which exhibit increasingly complicated oscillations. Most of these reaction schemes are in three dimensions and are accompanied by *rate equations* in discussing the behavior of the system.

Unlike the two dimensional systems, in three dimensional examples a number of complex solutions can be obtained. These will be illustrated by abstract models.

Significance of these models is that the complicated solutions are shown to exist even for simple nonlinear dynamic systems. Recently some of these models have been applied to explain the oscillations in experimental systems, e.g. Olsen and Degn (1977).

Model (1975-1)

Reference: (1975) Rossler.

The abstract reaction proposed is governed by the kinetic scheme as follows:

Reaction Scheme:

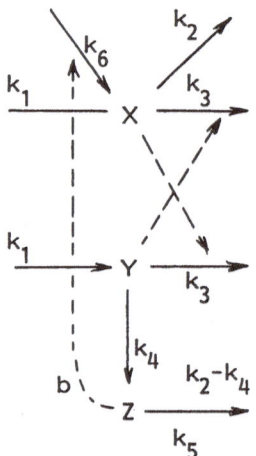

The dotted arrows indicate *catalytic rate controls*. The rate equations are given as follows:

42

Differential Equations:

$$dx/dt = -k_2 x - k_3 y \frac{x}{K + x} + k_1 + k_6 z$$

$$dy/dt = -k_2 y - k_3 x \frac{y}{K + x} + k_1 + b$$

$$dz/dt = k_4 y - k_5 z .$$

Oscillatory Solutions

In this 3-dimensional system a *stable limit cycle* is obtained as an oscillating solution.

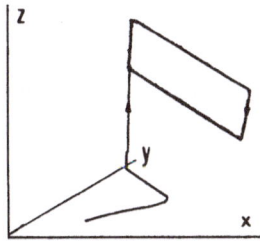

Fig. III.33. Limit cycle solution (After Rossler (1975))

Model (1976-1)

Reference: Rossler (1975).

Reaction Scheme

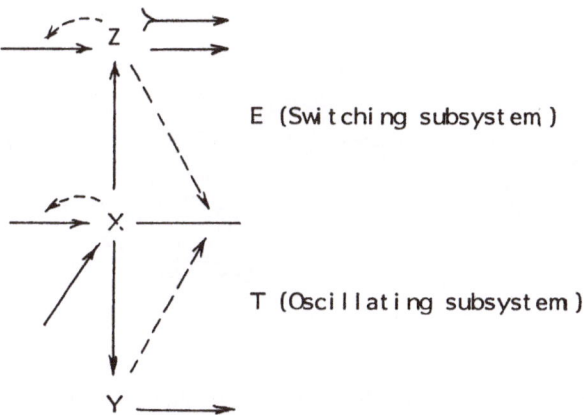

E (Switching subsystem)

T (Oscillating subsystem)

43

Okan Gurel and Demet Gurel

Differential Equations

$$dx/dt = k_1 + k_2 x - \frac{x(k_3 y)}{(k + x)} - \frac{x(k_4 z)}{(k + x)}$$

$$dy/dt = k_5 x - k_6 y$$

$$dz/dt = k_7 x + k_8 z - k_9 c^2 - \frac{k_{10} z}{c + K'}.$$

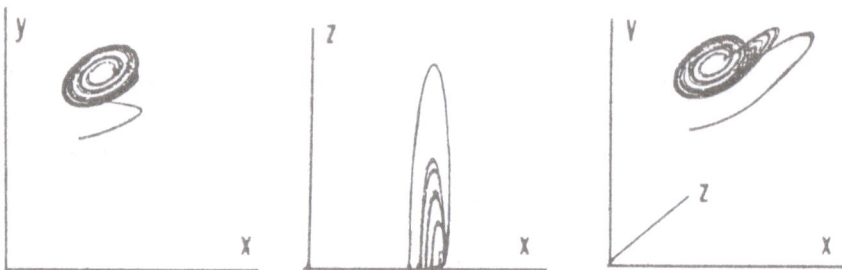

Fig. III.34. Projections on **a)** x–y, **b)** x–z, planes and the three dimensional plot of an oscillating solution. (After Rossler (1976-1))

Model (1976-2)

References: Rossler (1976-2), Gurel (1977).

Differential Equations:

$$dx/dt = -y - z$$

$$dy/dt = x + ay$$

$$dz/dt = -cz + xz + b.$$

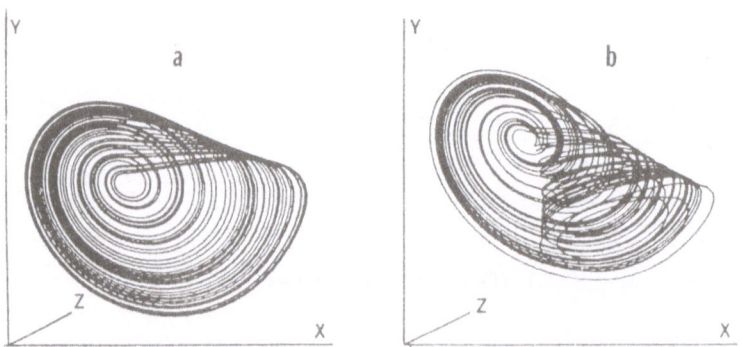

Fig. III.35. a Spiral solution (After Rossler (1976-2)), **b** Screw solution (After Rossler (1974-4, 5))

44

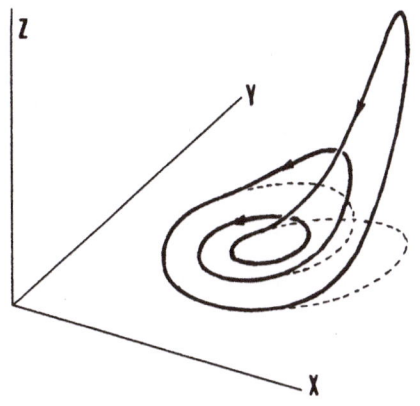

Fig. III.36. Limit Bundle (After Gurel (1977))

Model (1976-3)

Reference: Rossler (1976-3).

Differential Equations:

$$dx/dt = k_1' x - k_2 \frac{xy}{x + K} + k_5$$

$$dy/dt = k_3 x - k_4 y - Dy + Dy'$$

$$dx'/dt = k_3 x' + Dy - k_4 y' - Dy'$$

$$dy'/dt = k_1' x' - k_2 \frac{x' y'}{x' + K} + k_5.$$

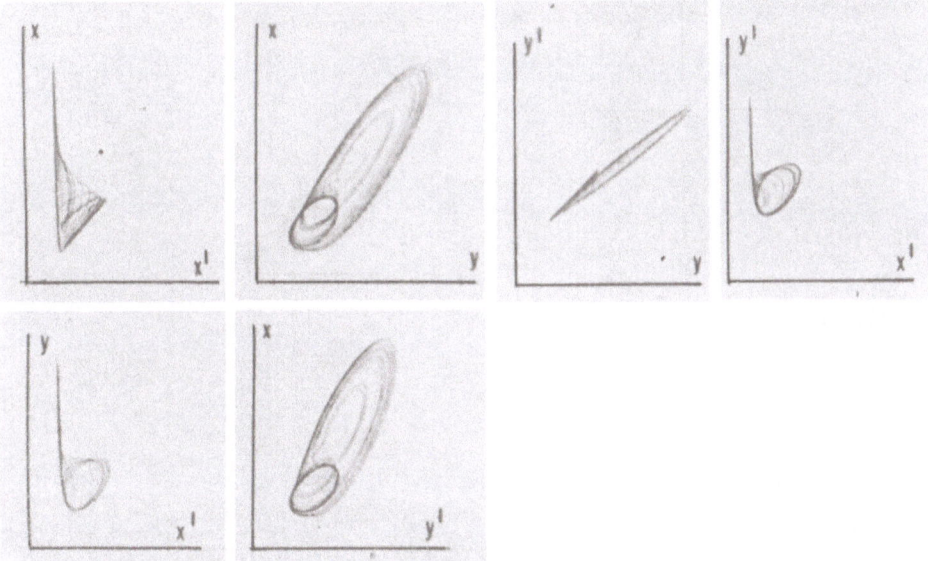

Fig. III.37. Six different projections of the oscillating solution. (After Rossler (1976-3))

Okan Gurel and Demet Gurel

Model (1976-4)

References: Rossler (1976-4), Gurel (1979-1), Gurel (1981).

Differential Equations:

$$dx/dt = x - z - xy$$
$$dy/dt = -ay + x^2$$
$$dz/dt = bx - cz + d.$$

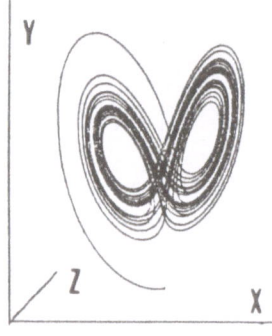

Fig. III.38. An exploded point type oscillating solution. (After Rossler (1976-4))

Model (1977-1)

References: Rossler (1977-1).

Reaction Scheme:

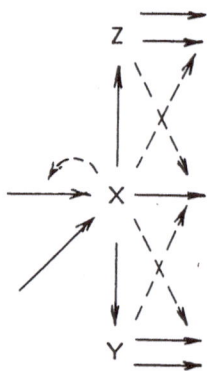

Differential Equations

$$dx/dt = ax - \frac{yx}{x + K_1} + c$$

$$dy/dt = x - by - \frac{zy}{y + K_2}$$

$$dz/dt = dx - e(x^2 + f)\frac{z}{z + K_3}$$

46

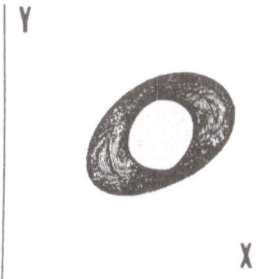

Fig. III.39. (After Rossler (1977-1))

Model (1977-2)

Reference: Rossler (1977-2, 3), Gurel (1973).

Spiral Type Chaos

Reaction Scheme:

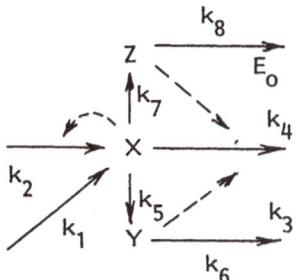

Differential Equations:

$$dx/dt = k_1 + k_2'x - (k_3y + k_4z)\, x/(x + K)$$
$$dy/dt = k_5x - k_6y$$
$$dz/dt = k_7x - k_8'z/(z + K').$$

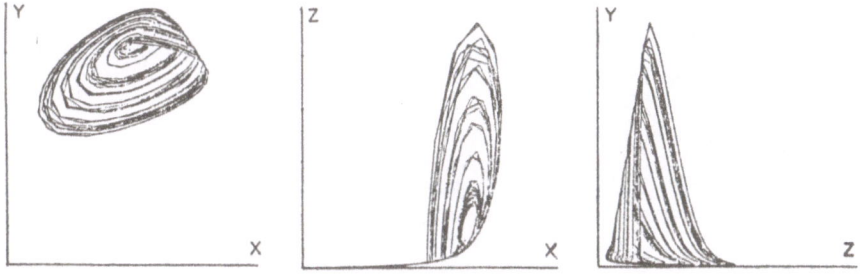

Fig. III.40. Spiral chaos (After Rossler (1977-2, 3))

Okan Gurel and Demet Gurel

Screw-Type Chaos

Reaction Scheme:

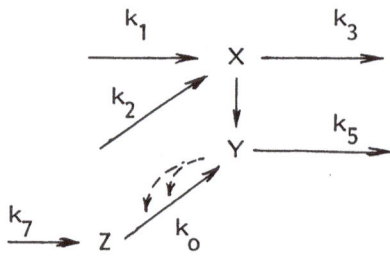

Differential Equations:

$$dx/dt = k_1 + k_2'x - k_3yx/(x + K)$$
$$dy/dt = k_4x - k_5y + k_6zy^2$$
$$dz/dt = k_7 - k_6zy^2 .$$

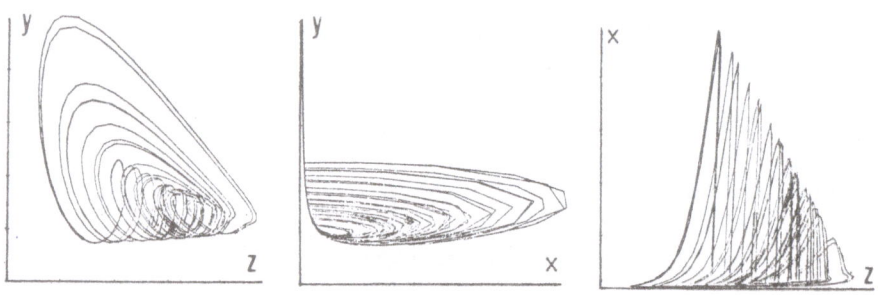

Fig. III.41. Screw-type chaos (After Rossler (1977-2, 3))

Model (1977-4):

References: Rossler (1977-4, 5).

Differential Equations:

$$dy/dt = -y - z$$
$$dy/dt = x$$
$$dz/dt = a(1 - x^2) - bz .$$

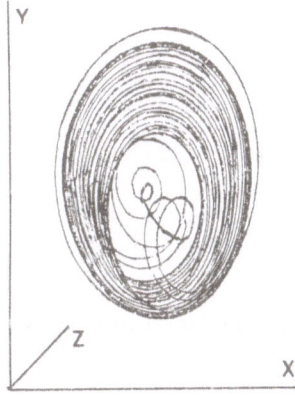

Fig. III.42. Screw-type chaos (After Rossler (1977-4))

Model (1977-5)

Differential Equations:

$$dx/dt = -ax - y(1 - x^2)$$
$$dy/dt = m(y + bz - cz)$$
$$dz/dt = m(x + dy - ez).$$

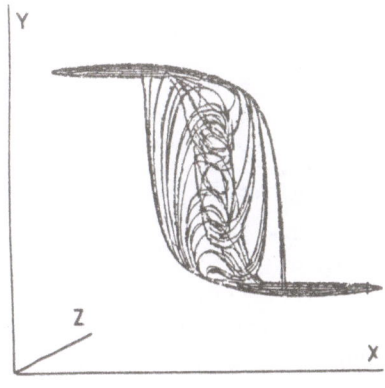

Fig. III.43. Screw-type chaos (After Rossler (1977-4, 5))

Model (1979-1)

Reference: Gurel and Rossler (1979).

Differential Equations:

$$dx/dt = x - ay - xz$$
$$dy/dt = bx + y - yz$$
$$dz/dt = x^2 + y^2 - \frac{z}{z + D}.$$

Okan Gurel and Demet Gurel

Fig. III.44. Creation of a torus (After Gurel and Rossler (1979))

Model (1979-2)

References: Rossler (1979-2, 3, 4, 5, 6, 7).

Differential Equations:

$$dx/dt = -y + ax - bz$$
$$dy/dt = x + c$$
$$dz/dt = d(1 - z^2)(x + z) - z.$$

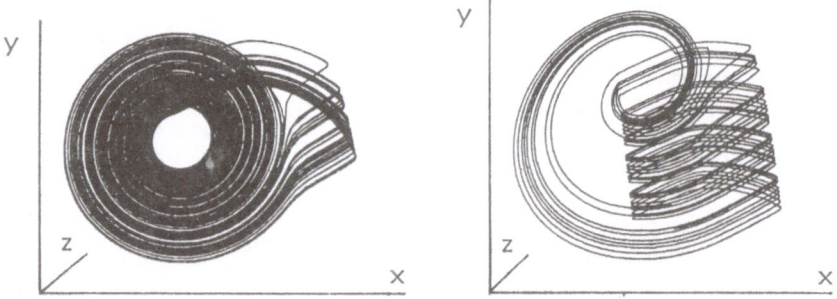

Fig. III.45. Spiral and Screw Type chaos. (After Rossler (1979))

Model (1979-3)

Reference: Rossler (1979).

Differential Equations:

$$dx/dt = x - xy - z$$
$$dy/dt = x^2 - ay$$
$$dz/dt = b(cx - z).$$

Fig. III.46. Inverted spiral-plus-saddle type chaos. (Lorenzian Chaos), (After Rossler (1979))

Model (1979-4)

Reference: Rossler (1979).

Differential Equations:

$$dx/dt = -xy - ax - z$$
$$dy/dt = -x + by + cz$$
$$dz/dt = d + exz + fz .$$

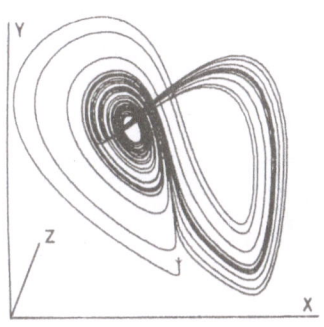

Fig. III.47. Noninverted spiral-plus-saddle type chaos (After Rossler (1979-2))

Model (1979-5)

Reference: Rossler (1979).

Differential Equations:

$$dx/dt = -y - z$$
$$dy/dt = x$$
$$dz/dt = a(y - y^2) - bz .$$

Okan Gurel and Demet Gurel

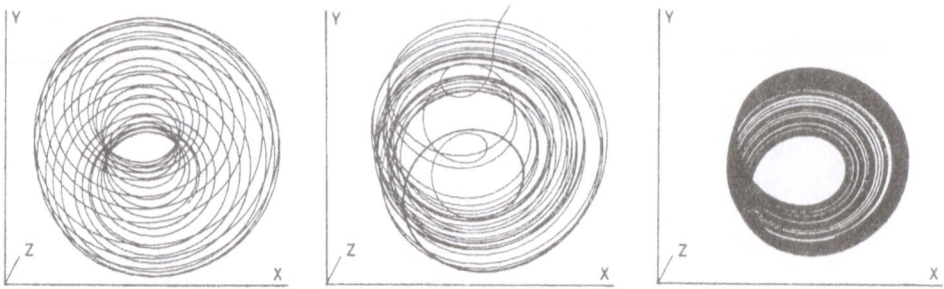

Fig. III.48. Toroidal behavior (After Rossler (1979))

Model (1979-6)

Reference: Rossler (1979).

Differential Equations:

$$\mathrm{d}x/\mathrm{d}t = -y - z - w$$
$$\mathrm{d}y/\mathrm{d}t = x$$
$$\mathrm{d}z/\mathrm{d}t = a(y - y^2) - bz$$
$$\mathrm{d}w/\mathrm{d}t = c(z/2 - z^2) - dw \,.$$

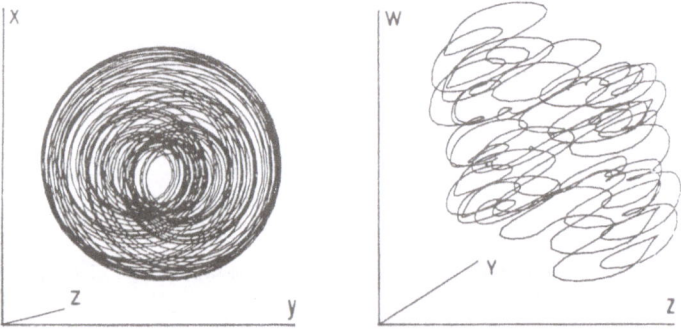

Fig. III.49. Hypertoroidal behavior (After Rossler (1979))

Model (1980)

Reference: Williamowski and Rossler (1980).

Differential Equations:

$$\mathrm{d}x/\mathrm{d}t = x(a_1 - k_1^* x - z - y) + k_2^* y^2 + a_3$$
$$\mathrm{d}y/\mathrm{d}t = y(x - k_2^* y - a_5) + a_2$$
$$\mathrm{d}z/\mathrm{d}t = z(a_4 - x - k_5^* z) + a_3 \,.$$

Reaction Scheme:

$$A1 + X \underset{k1^*}{\overset{k1}{\rightleftarrows}} 2X$$

$$X + Y \rightleftarrows 2Y$$

$$A5 + Y \rightleftarrows A2$$

$$X + Y \rightleftarrows A3$$

$$A4 + Z \rightleftarrows 2Z .$$

For constant A's the resulting figure is a limit cycle.

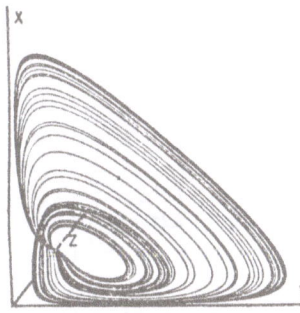

Fig. III.50. Chaotic oscillations in Model (1980). (After Rossler (1980))

Table III. Classification of Abstract Models by Right-Hand Sides

Model	Constant	x	y	z	x^2	y^2	z^2	A	B	C	$D1$
1975-1	$k1$ $k1^*$	$-k2$	$-k2$ $k4$	$k6$ $-k5$				$-k3$	$-k3$		
1976-1	$k1$	$k2$ $k5$ $k7$	$-k6$	$k8$			$-k9$	$-k3$		$-k4$	$-k10$

where $A = xy/(k + x)$, $B = xy/(k + y)$, $C = xz/(k + x)$, $D1 = z/(k^* + z)$.

Model	Constant	x	y	z	xy	xz	yz	x^2	y^2	z^2	A_1
1976-2	 b	 1	1 a	-1 $-c$	-1	1					
1976-4	 d	1 $-a$	b	-1 $-c$	-1			1			

Table III. (continued)

Model	Constant	x	y	z	xy	xz	yz	x^2	y^2	z^2	A_1
1977-4				−1	−1						
	1										
	a			−b	−a						
1979-1		1	−a			−1			1		
	b	1				−1		1			
				c					1		−1
1979-3		1		−1	−1						
			−a	−b					1		
		bc									
1979-4			−a	−1	−1						
			−1	b	c						
	d		f				e				
1979-5			−1	−1							
		1									
			a	−b					−a		
1981	a3	a1			−1	−1		−k1*	k2*		
	a2		−a5			1			−k2*		
	a3			a4		−1				−k5*	

where $A_1 = z/(z + d)$

Model	Constant	x	y	z	A	B	C	D	E	F
1977-1	c	a			−1					
		1	−b					−1		
		d							E1	
1977-2	k1	k2′				−k3				
		k5	−k6					−k4		
		k7							E2	
1977-3	k1	k2′				−k3				F1
		k4	−k5							−F1
	k7									
1977-5	−a		−1							F2
		m1	m2	−m3						
		m4	m5	−m6						

where $D = yz/(k_1 + y)$, $E1 = e(x^2 + f)\, z/(z + k_3)$, $F1 = zy^2$
$E2 = -k8\, z/(z + k)$, $F2 = yx^2$.

Model	Constant	x	y	z	xz^2	z^3
1979-2		a	−1	−b		
	c	1				
		e		e−1	−1	−1

Table III. (continued)

Model	Constant	x	y	z	w	x^2	y^2	z^2	w^2	G	H
1976-3	k_5	k_1^*								$-k_2$	
		k_3	$(-k_4 -D)$		D						
			D	k_3	$(-k_4 -D)$						
	k_5			k_1^*							$-k_2$
1979-6			-1	-1	-1						
		1									
			a	$-b$			$-a$				
				$c/2$	$-d$			$-c$			

where $G = xy/(k + x)$, $H = zw/(k + z)$.

K Uncatalyzed Reactions

References: Koros and Orban (1978), Orban and Koros (1978), Koros and Orban (1980).

Koros and Orban have studied (1978-1, 2, 1980) the uncatalyzed bromination of aromatic compounds and observed oscillations.

L Tableau of Oscillatory Reactions

To summarize the reactions discussed above with their models and oscillatory solutions Table IV is prepared. For each reaction (Column 1) the references of experimental motivation for studying oscillations (Column 2), and the references for mathematical models associated with each reaction (Column 3) are given. The oscillatory solutions are listed in Column 4.

Table IV. Examples of Oscillatory Reactions and Models

Class of Reaction	Experimental Motivation	Mathematical Model	Class of Oscills.
Bray-Liebhafsky	Auger (1911) Bray (1921)	Matsuzaki et al. (1974) Edelson and Noyes (1974)	1 LC 1 LC
• iodine • hydrogen peroxide • feedback			
Briggs-Rauscher • iodine	Briggs and Rauscher (1973)	Boissonade (1976)	1 LC
• manganous ion • malonic acid to Bray system • B-L variation			

Table IV. (continued)

Class of Reaction	Experimental Motivation	Mathematical Model	Class of Oscills.
Belousov-Zhabotinskii	Belousov (1958)	Noyes and Field (1972)	1 LC
	Zhabotinskii (1964)	Rossler and Wegmann (1978)	Chaos
• Acidic bromate		Wegmann and Rossler (1978)	Chaos
• oxidation			
• catalyzed by cerous and manganous ion			
CSTR (Continuous Stirred Tank Reactor)		Bilous and Amundson (1955)	1 LC
		Aris and Amundson (1958)	M LC's
Solid catalyzed reactions			
N$_2$O decomposition	Hugo (1968)		
H$_2$ oxidation	Beusch et al. (1970)		
CO oxidation	Hugo (1970)	Sheintuch (1977)	LC
	Beusch et al. (1970)		M LC's
Glycolysis	Duysen and Amesz (1957)	Higgins (1964)	1 LC
Glycolytic		Sel'kov (1968)	M LC's
enzyme system		Goldbeter (1974)	1 LC
• homogeneous		• feedback	
• enzyme			
• phosphofructokinase			
• NADH			
• feedback			
Peroxidase Reactions	Yamazaki et al. (1965)		
	Degn (1968)		
Na$_2$S$_2$O$_4$ decomposition	Rinker et al. (1965)		
	DePon and Mason (1974)		
Bimolecular	—	Lefever (1968)	1 LC
Abstract Chemical Models		Rossler (1970s)	Chaos
Uncatalyzed Reactions	Koros and Orban (1978)		

In this table 1 LC stands for a single limit cycle, and M LC indicates multiple limit cycles.

IV Characteristics of Oscillatory Systems

Oscillations are special solutions of a dynamic system. These solutions may not only exist but also exhibit certain characteristics which would differentiate them from each other. One of the factors entering into this differentiation is the *dimension* of the system. Below we discuss the different forms of basic oscillatory solutions for systems with various dimensions. It is shown that as the dimension increases the possibilities of richer oscillating solutions also increase.

Another element contributing to the existence and types of oscillatory solutions is the *nonlinearity* of the system. A classification of known chemical models based on their nonlinearities is given below. It is shown that various models discussed in the literature by different investigators may be grouped in special classes.

A Dimension of a System

The number of chemical reactants determines the dimension of a chemical system. The minimum number is necessarily one, however the interaction between the elements is interesting thus systems with at least two reactants is relevant in the context of the present paper. In mathematical terms, a solution is a line represented by $x = x(t)$ as a function of time element. Representation of this solution has been in basically two different forms which will be discussed below:

A solution $x(t)$, is a *line* in (x, t)-space. Its representation in the entire solution space with dimension greater than one is possible by embedding the line into the space. Because of this geometry solutions appear in different forms in spaces with different dimensions. As a consequence of increasing dimension of the solution space we have various possibilities.

The Systems with Dimension 1: (Boundedness of Solution)

Traditionally oscillations have been observed and presented in a one dimensional geometry, a line. In mathematical terms this is $x = x(t)$ as shown in Fig. IV.1.

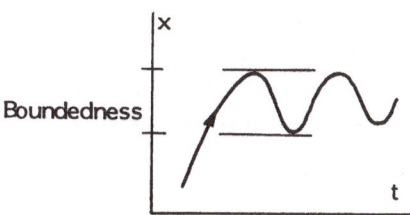

Fig. IV.1. Oscillatory solution in one-dimensional representation

Here t is the time element of the dynamic (kinetic) system, x is the variable, concentration of a chemical element of the reaction.

This representation is suitable if one is interested particularly in both amplitude (measured along the x-axis), and period (measured along the t-axis), both being (quantitative characteristics). Furthermore it may be used for more than one variable, e.g. x and y by plotting multiple graphs for (x, t), (y, t), ... The only qualitative information we get is that x is *bounded*.

57

The Systems with Dimension 2: (Boundedness and Closedness)

Since the time of Poincare elimination *t* from the presentation and focusing on the variables only has been accepted as an alternative for representation. In two dimensional systems this is called the *phase plane*, Fig. IV.2.

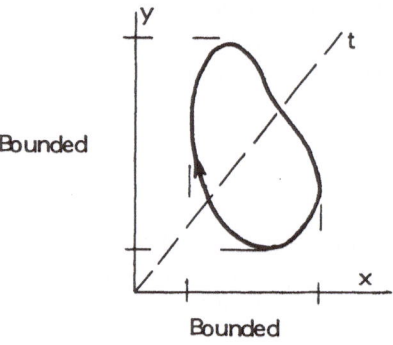

Fig. IV.2. A limit cycle, a bounded solution in two dimensional plane

In fact *t*-axis still exists, but only the projection of the system dynamics $x(t)$, $y(t)$ onto (x, y)-plane is shown. The time element is no longer explicit, thus measuring periods is not possible. However, one can measure amplitudes clearly. Moreover, this representation is suitable more for visualizing (qualitative characteristics) of the system than for measuring quantitative characteristics.

In the two dimensional space, the plane, the most sriking solution is a *closed* curve, named by Poincare as the *limit cycle*. A limit cycle is bounded as well as closed, and it is the oscillating solution of the dynamic system under consideration.

One of the significant qualitative characteristics of two dimensional solutions could be related to the concept of *convexity* of the area bordered by the closed solution, the limit cycle. One can easily visualize the $x(t)$ and $y(t)$ representation in which both *x* and *y* exhibit smooth oscillations for convex, and oscillations with multiple periods for concave limit cycles, Fig. IV.3. Clearly one example of the latter oscillations may well be "double oscillations" discussed in Section III.

Both convex and concave limit cycles can be stretched to a circle, thus the limit cycle is the only periodic solution in two dimensional systems. A limit cycle is *bounded* in both x and y directions. Moreover by definition it is a *closed* curve.

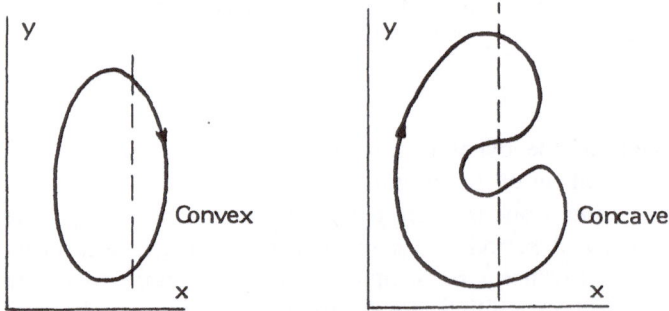

Fig. IV.3. Convex and concave areas bordered by a limit cycle

We should also note here that in two-dimensional examples there can be *multiple cycles*, e.g. in some Sel'kov examples of glycolysis.

Examples in dimension 2:

- CSTR models
- Bimolecular model
- Glycolytic models.

The Systems with Dimension 3: (Boundedness, Closedness or Non-closedness)

In three dimensional systems, both qualitative and quantitative representation may be used. However, the qualitative representations in *phase space* reveal characteristics about the system which can not be discovered by quantitative evaluation only.

The characteristics of a limit cycle in three dimensional space are somewhat different from that in two dimensional case. The generic possibilities are shown in Fig. IV.4.

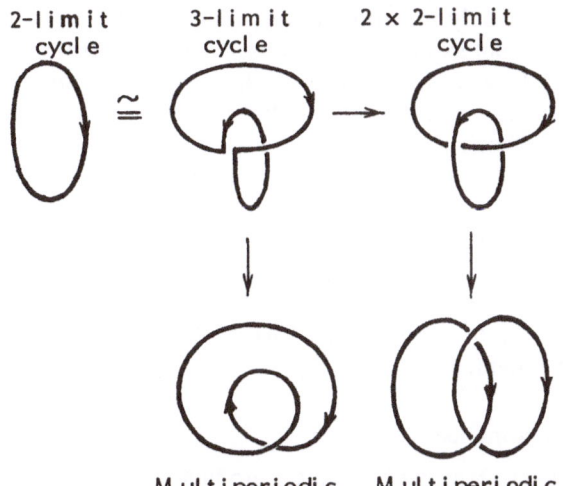

2-limit cycle 3-limit cycle 2 x 2-limit cycle

Multiperiodic limit cycle Multiperiodic Multi limit cycles

Fig. IV.4. Generic limit cycles in three dimensional space. Here, 2-limit cycle is a limit cycle in the plane, 3-limit cycle is a limit cycle in 3-dimensional space

In these systems we still have the basic qualitative characteristic of *boundedness* for *closedness* restriction may be relaxed to open up new possibilities. Referring to the functional dimension (*f*-dimension) of the mathematical solutions, see Gurel (1981), the *bounded* solutions in three dimensional space may have three different forms:

- zero *f*-dimensional: (open) exploded point (EP)
- one *f*-dimensional: (closed) limit cycle (LC)
- two *f*-dimensional: (open) limit surface (LS)

Examples in dimension 3:

- Briggs-Rauscher
- Oxidation of malonic acid
- Abstract Models of Rossler

The Systems with Dimension n greater than 3

It is quite obvious that already in three dimensional systems the monotony of limit cycles is replaced by rich possibilities of periodic solutions. In higher dimensional problems more intricate solutions may also be encountered. These specifically are *hypersurfaces* with different characteristics.

Examples in dimension 4:

- Abstract models (1976-3)
- Abstract models (1979-6)

B Nonlinearity of a System

Nonlinearity of a system is reflected in the right-hand sides of the differential equations forming the model corresponding to a reaction kinetics. These differential equations are the *rate equations* representing the mathematical model of a given chemical scheme. These equations not only incorporate the *rate constants* (k's) but also *how* each of the *reacting elements* enters into one or more reactions. Depending on the form of these interactions between the elements, the *nonlinearity* of the system can be determined. In turn, this nonlinearity leads to particular types of solutions with different oscillatory as well as nonoscillatory characteristics.

Nonlinearity of a system also arises from the interaction of an element being a product as well as a controlling factor of any one of the reaction steps preceding this product. If a nonlinearity results, this type of nonlinearity is called a *feedback* nonlinearity.

Moreover, the interactions of the elements would also determine the mathematical form of the right-hand side. The forms of nonlinearities used in known models are discussed below and examples are provided.

1 Feedback Mechanism Leading to Nonlinearity

In certain reactions nonlinearity stems from the fact that a product feeds back to a particular stage of a reaction and either *activates* or *inhibits* the reaction causing a form of nonlinearity in the reaction scheme. Higgins (1964) has formalized these concepts in his paper.

The feedback phenomenon has been recognized and referred to in the chemical literature. For example, specifically we can list the following examples from the literature:

> Bray-Liebhafsky reaction: Bray (1921)
> Glycolysis: Higgins (1964).
> Catalytic Oxidation of Hydrogen: Belyaev (1973)
> See also the review articles: Gurel (1972), Franck (1979)

2 Product Nonlinearity

One class of nonlinearity in mathematical models involves the one produced by a *product* of variables, such as x^2, x^3, xy, yz^2, etc. This nonlinearity may introduce considerable variations of oscillating solutions. Examples from the literature with product nonlinearity are listed below:

Bimolecular model: Lefever (1968)
Briggs-Rauscher: Boissonade (1976)
Belousov-Zhabotinskii: Field and Noyes (1974)
Oxidation of CO: Yang (1974)
Oxidation of $Na_2S_2O_4$: De Poy and Mason (1974)
Abstract models: Rossler (1976-2 and 4), (1977-4 and 5), (1979-2 through), (1980)

3 Exponential Nonlinearity

In some models, particularly those from chemical engineering applications, the nonlinearity in the heat balance and mass balance equations contain *exponential* terms. Examples of exponential nonlinearity are,

. CSTR: Bilous and Amundson (1955)
CSTR: Aris and Amundson (1958)
CSTR: Gurel and Lapidus (1965)
Catalytic decomposition of N_2O: Eckert (1977)
Catalytic oxidation of hydrogen: Pikios and Luus (1977)

4 Rational Nonlinearity

In a number of models, nonlinearity is introduced as rational terms. Michaelis-Menten reaction kinetics also leads to this type of nonlinearity. Some examples from the literature are given below.

Glycolysis: Higgins (1964)
Glycolysis: Sel'kov (1968)
Abstract models: Rossler (those not included in Product Nonlinearity section above.)

C Parameters of the System

In the case of dynamic systems exhibiting various forms of solutions, such as stationary points as well as oscillating solutions, in addition to the variables and their behavior there is an important set of entities known as *parameters*. These parameters may well be "variables", however their role is clearly different from that of the variables. Parameters appear as either coefficients of the terms of the dynamic equations, or a power of a variable or constants entering into the equations. Not necessarily all but some of these parameters, as they vary, affect the solutions of the system, thus for certain intervals of the parameter values, the dynamic system may exhibit different behaviors.

Usually *reaction rates* are the most common parameters in chemically reacting systems. Models discussed in the literature may be studied from the point of view of parameters. Particularly, in the bifurcation analysis of reaction models we refer to variations in parameters, and based on these variations the appearance and the disappearance of oscillating solutions may be predicted.

Among the investigators studying models for various parameter values are Sel'kov and Rossler, see models in Sections III.F and III.J.

D Number of Solutions

In oscillatory systems, due to the nonlinearity and the variations in parameters multiple solutions, both oscillatory and nonoscillatory simultaneously may appear. Multiple stationary solutions have been discussed by Bilous and Amundson (1955), Aris and Amundson (1958), see Section III.D. Furthermore, multiple oscillating solutions have been extensively studied by Sel'kov and his collaborators, see Section III.F.

V Global Analysis

Global analysis is an approach to analyze dynamic systems such as chemical reactions, by referring to the global characteristics of the behavior of the system. Unlike the local characteristics of a singular point, global characteristics apply to a larger domain, in fact to the entire solution space. One of the important concepts to understand is the qualitative versus quantitative thinking. Historically experimentalists have observed one variable out of the entire set of possible variables interacting in reaction. Thus quantitative analysis prevailed. As the theoretical approaches advanced, methods to observe system globally became feasible, and *global analysis* helped improve our understanding of complex systems such as chemical reactions. One of the tools of global analysis, the bifurcation theory emerged as a significant tool in the global analysis of oscillating chemical systems.

A Qualitative versus Quantitative Analysis

Clearly, *oscillations* are the main characteristic of the oscillating systems. Although in the past the *quantitative* aspects of these oscillations, such as their *period* and *amplitude*, were significant and the only information sought, as the new discoveries about the systems made, the *qualitative* aspects of oscillations, such as *multiple periodicity, nonperiodicity, and chaotic oscillations* gained importance over the quantitative elements. It is difficult to deduce a conclusion on the behavior of an oscillatory solution based solely on quantitative elements such as its periodicity and amplitude, because these elements appear on the *projection* of (the solution space vs time) onto (one-variable vs time) space. Thus the global "picture" of this dimension together with the remaining elements of the solution space, based on the entire solution space, is lost. However, if a *global* approach is used one can qualitatively observe the behavior of an oscillating solution in a global setting with all the available information about the system.

B Bifurcation as a Tool for Detecting Oscillations

Bifurcation phenomenon is a mathematical concept introduced by Poincare, see Poincare (1885). Although special, the phenomenon has quite general applications, (Gurel 1979-2). Since the theory is the study of creation of solutions as the parameters of the system vary the appearance and the disappearance of oscillatory solutions can also be studied by the theory of bifurcations.

As pointed out in (Gurel 1972), Higgins in his 1964 study observed the appearance and the disappearance of limit cycles in glycolysis implying the possibility of bifurcations, however neither mentioning bifurcations specifically, nor emphasizing this nonlinear phenomenon.

On the other hand as early as 1958, Aris and Amundson (1958) discussed the role of bifurcations in the appearance of limit cycle solutions of the CSTR reactions. A thorough bifurcation analysis of these equations were completed by Uppal et al. only in 1974.

Among the existing models, one representing the reaction discovered by Belousov has also attracted bifurcation studies. However, more interesting bifurcation studies are based on the models constructed by Rossler, due to the fact that these systems possess a rich set of solutions both oscillatory and nonoscillatory as well as periodic and nonperiodic oscillating solutions, Gurel (1977), Gurel and Rossler (1979).

1 Parameters of the System

In bifurcation analysis, some of the parameters of the system equations, see discussion in Section IV.C, are the elements affecting the behavioral change in solution space. It is necessary to determine which one of these parameters are the *bifurcation parameters*. Variations in bifurcation parameters would result in various possibilities of solutions, such as oscillatory solutions, multiple solutions, etc. See, e.g. the simple analysis in Section III.I.

2 Stability of Solutions

An important concept in the dynamical systems is the stability property of the solutions. In an experimental environment only those solutions which are *stable* can be *observed*. In a system there can be *multiple* stable solutions, such that depending on the *initial values* of variables in a dynamical system the system would move towards one or another stable solution.

Prior to recognizing the possibility of multiple solutions, investigators referred to the "observed" stable solutions whether a singular point or an oscillating solution as the "unique" stable solution of the system. In many cases this would be a correct conclusion, such as in Bimolecular model (1968). Moreover, investigators erroneously "assumed" that if a system has a stable singular solution under one experimental environment and a stable limit cycle under another experimental environment, in order to observe the stable limit cycle the original stable singular point "must" become unstable. In many cases this "assumption" is also true, however there are examples where both the singular point and the limit cycle are stable, separated by an unstable limit cycle. In fact, the bifurcation theory can show that all the possible bifurcations may lead to a rich set of various combinations of stable and unstable, oscillatory and nonoscillatory solutions. In the chemical literature this fact was clearly emphasized, Gurel (1979-1), and ample examples are known and discussed in the literature on bifurcations. In fact as late as in 1980, there are still research papers appearing in the chemical literature based solely on the "rediscovery" of this fact by the investigators, e.g. Tockstein and Komers (1980).

3 Number of Solutions

In the case of nonlinear systems one of the interesting features of a system is that *multiple solutions* may occur. In the case of "bifurcating" systems there are possible intervals of the parameters in which the system exhibits only one singular solution which for all practical purposes is a *stable singular point*. For some parameter intervals other than the above, bifurcation in the system is realized, and this singular point bifurcates into multiple solutions. For example in the simple two-dimensional system of Lefever (1968), the only possibility is the case of an unstable singular point surrounded by a *stable limit cycle* simultaneously bifurcating from the original stable singular point.

Multiple solutions more complex than this are also present in the literature. For example for two-dimensional systems Aris and Amundson (1958) discussed multiple singular solutions as well as multiple limit cycles appearing simultaneously.

In three-dimensional systems there can be even more interesting examples of multiple solutions. In addition, some of these systems (models) may exhibit different sets of multiple solutions for different intervals of parameters. The most complete analysis of such a case is given in Gurel and Rossler (1979).

4 Detection of Oscillatory Solutions

As mentioned above the bifurcation phenomenon may be utilized to analyze oscillatory systems and their models. In a chemical reaction, observation of oscillations in the concentration of a reactant would imply further possibilities of oscillating concentrations. Experimental observation of these oscillations automatically imply the stability of the corresponding solution of the mathematical model. Therefore, from the experimental findings, a reaction scheme and a mathematical model may be constructed. Assuming that the scheme and the model are the "true" representations of the chemical experiment, it is clear that a stable oscillatory solution of the mathematical model would correspond to the observed oscillations. Moreover, if the model has *parameters*, as the *stability changes* and *multiple solutions* occur the system *may be* an example of *bifurcation*, thus the newly found stable oscillatory solutions different from the previously observed ones would also be *feasible*. Further experiments may be designed to verify these mathematical solutions, thus solidifying the validity of the models and benefiting from the theory.

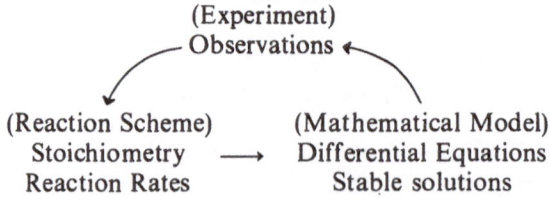

(Experiment)
Observations

(Reaction Scheme)   (Mathematical Model)
Stoichiometry   ⟶   Differential Equations
Reaction Rates   Stable solutions

Oscillations being *qualitative* characteristics of a dynamic system, *global analysis* in particular via *bifurcation theory* becomes a primary tool of detecting such characteristics of a dynamic system. On the other hand, *quantitative* aspects of oscillations such as their periods, and amplitudes would supply further *detailed* information about a particular oscillation.

In this review we summarized the oscillatory chemical reactions reported in the literature. As collected in Table IV, the number of oscillating reactions reported is limited. Some of the most studied reactions have been observed and reported some time ago. Subsequently reaction schemes have been proposed to elucidate the reaction mechanism. In some cases also a mathematical model has been constructed to show that the model possesses a stable oscillatory solution. Further, closing the loop by going back to the design of an experiment exhibiting "other" oscillatory solutions remains to be explored. Abstract models of Rossler appearing in the literature since 1975 show the possibilities of various stable solutions, thus providing experimental chemist with a rich glossary of oscillations. In fact, there is a clear attempt by Olsen and Degn (1977) to find a chemical reaction where chaotic oscillations predicted by abstract models are observed, Figure III.31.

VI Classification of Oscillations

Classification of oscillations in chemical systems will be made in terms of three groupings; chemical reactions, chemical elements, and oscillatory mathematical solutions. In the first grouping an enumeration of different "reactions", are considered. In the second, a listing of those chemical elements which take part in oscillatory reactions are considered. Finally, the oscillatory solutions of the mathematical model represented by the rate equations are classified on the basis of known examples.

A Types of Oscillatory Chemical Reactions

1. Catalyzed Reactions

a. Homogeneous Reactions

- *Enzyme Catalyzed Reactions*
    a. *Glycolysis* (e.g., phosphofructokinase catalyzed)
    b. *Oxidation* of NADH by horseradish peroxidase
    c. *Oxidation* of NADH by lactoperoxidase
    d. *Iodination* of thyroglobulin by thyroid peroxidase

- *Metal Ion Catalyzed Reactions*
    a. *Oxidation* of malonic (citric) acid
    b. *Bromination* of organic acids with active methylene groups

- *Iodate Catalyzed Reactions*
    a. *Decomposition* of hydrogen peroxide

- *Autocatalytic Reactions*
    a. *Decomposition* of sodium dithionite

b. Heterogeneous Reactions

- *Solid Catalyzed Reactions*

    a. *Decomposition* of N_2O
    b. *Oxidation* of CO
    c. *Oxidation* of H_2

2. Uncatalyzed Reactions
    a) *Oxidation* of aromatic compounds with bromate

B Types of Elements of Oscillatory Chemical Reactions

1. Catalysts

- *Enzymes*
    a. Peroxidases
    b. Allosteric enzymes (phosphofructokinase)

- *Metal Ions*

    With strongly positive oxidation states e.g., cerium, manganese, etc.

- *Surface Catalysts*

    Nickel, platinum, etc.

2. Substrates
    - *Halogens* with positive oxidation states, e.g., iodine, bromine
    - CO
    - N_2
    - H_2

C Types of Oscillatory Mathematical Solutions

Oscillations of physicochemical elements such as temperature, concentrations of chemical elements, etc. correspond to oscillatory solutions of the dynamic equations (differential equations) of the system under consideration. As discussed above, various reactions possess different types of these oscillatory mathematical solutions. The well known *limit cycles* were first observed and named as such by Poincaré exactly a century ago. Although formulated by Poincaré in general terms, the other mathematical solutions have recently been explored and named by mathematicians, e.g. *attractors* and *exploded points* consequently their application to chemical systems are more recent than the limit cycles.

Limit Cycle

The limit cycle, particularly of two-dimensional systems, is the simplest form of the oscillatory solutions. As discussed in Section IV.A. in two- and higher-dimensional systems limit cycle solutions may occur. In the literature almost all oscillatory examples are some type of a limit cycle. An example of a simple limit cycle is the bimolecular model of Lefever. However, as the dimension of the system increases not only other oscillatory mathematical objects may appear, but also limit cycles may become complex. As shown above in Section IV.A., in three-dimensional systems complex limit cycles illustrated by generic examples, Fig. IV.4. may be obtained. In addition, "double oscillations" in examples of Beusch (1972), Figure III.13., Marek and Svobodova (1973), Vavilin, et al. (1973), Figure III.9., Dynnik and Sel'kov (1975) Fig. III.25., and Boissonade (1976), Fig. III.3 and 4., are possible projections of limit cycles with generic behavior shown as multi periodic limit cycle in Fig. IV.4.

Limit Bundle

The limit bundle shown in the abstract model (1976-2) is also generically a multi periodic limit cycle. In the abstract model (1974-4) linked limit cycles are also observed. These linked limit cycles are examples of the generic multi limit cycle shown in Fig. IV.4.

As the theoretical models reveal the possibilities of limit cycles with varying complexities, an awareness of these variations influences the experimental studies.

Attractors

Some of the oscillatory solutions, in particular those found in the abstract models of Rossler, are *attractors, moreover they are chaotic*. These mathematical solutions are interesting in that many oscillations observed experimentally are probably of this nature, and if studied with models encompassing the true behavior of the reaction they can be obtained theoretically. These chaotic attractors are illustrated by the examples given in Section III.I.

Some of these chaotic oscillations are obtained by constructing models to yield previously predicted geometries. Examples are the screw-type chaos, and toroidal chaos in Models (1977-3) and (1979-1, 2), respectively, obtained as chemical realization of coil-type oscillations introduced in Gurel (1973).

Using an opposite approach, that of starting from a mathematical solution and designing an experiment, Olsen and Degn (1972) showed that abstract models may lead to an understanding of the oscillatory chemical reactions exhibiting not only just limit cycle oscillations but also chaotic attractor-type oscillations.

Exploded Point

As it has been known for some time, in certain reactions, oscillations observed may not be modelled as periodic limit cycles. In fact, there may not be any period found during a reasonably long time interval. Mathematical solutions without a period may be observed in three-dimensional mathematical models, e.g. Model (1976-4). This solution was named generically as an exploded point, Gurel (1981).

It is quite likely that there might be numerous examples of chemical reactions which oscillate without any period thus forming condidates for exploded points. However, this remains to be verified.

VII State of the Art and Future Directions

In this article we reviewed the evolution of research in oscillatory chemical reactions particularly during the past decade. Characteristics of this evolution are as follows:

1. A number of chemical reactions were recognized in the early part of the century. Some of these have inspired chemists to reevaluate the oscillatory reactions.

2. Because of the advances in the mathematics of oscillatory solutions, chemical oscillations have been analyzed by taking these new mathematical concepts into consideration. Furthermore, some of the mathematical research has been motivated by the behavior of some chemical reactions.

3. New reaction schemes and new mathematical models have been introduced and extended to reactions known to exhibit oscillatory behavior. This evolution both in experimental and theoretical fields are summarized in Table IV.

As in the case of Olsen and Degn experiment (1977), a new mode of research as application of new mathematical findings to reevaluate chemical reactions clearly signaled a turning trend in research on chemical oscillations.

In the near future it can be expected that explanation of behavior of even complex chemical reactions will be attempted and types of oscillations will be more systematically classified. In this context, some of the previous work will probably be reevaluated. Since the role of chemical oscillations is clearly related to the biological systems via enzyme kinetics, chemical reaction studies will be centered no longer around the stationary states but the oscillatory solutions, both stable and unstable.

The past decade where an explosion of interest in oscillations has been witnessed will remain as the renaissance period in the field of research in oscillatory chemical reactions.

VIII References

References are listed in alphabetical order. The notations on the left hand column are (Section, Part). Review articles are denoted by an R and the number used in Table II. Books are shown with letter B.

(III D) 1958 Aris, R., Amundson, N. R.: An Analysis of Chemical Reactor Stability and Control. II. The Evolution of Proportional Control, Chem. Eng. Sci. vol. 7, 132–147

(III A) 1911 Auger, V.: Action de l'eauxoxygenee sur les composes oxygenes de l'iode, Compt. rendus. vol. 153, 1005–1007

(III C) 1959 Belousov, B. P.: Sb. Ref. Radiats. Med. 1958, Medzig (Moscow) p. 145. (1958 Collection of Abstracts on Radiation Medicine)

(III E) 1973 Belyaev, V. D., Slin'ko, M. M., Timoshenko, V. I., Slin'ko, M. G.: Generation of Auto-Oscillations in the Hydrogen Reaction on Nickel, Kinetics and Catalysis, vol. 14, 708–709

(III E) 1972 Beusch, H., Fieguth, P., Wicke, E.: Thermisch und kinetisch verursachte Instabilitäten im Reaktionsverhalten einzelner Katalysatorkorner, Chem. Ing. Tech. vol. 44, 445–451
See also (1972) Unstable Behavior of Chemical Reactions at Single Catalyst Particles, In: Chemical Reaction Engineering Reviews, Advances in Chemistry Series (ed. Gould, R. F.) 109, American Chemical Society, Washington, D.C. 615–621.

(III D) 1955 Bilous, O., Amundson, N. R.: Chemical Reactor Stability and Sensitivity, A. I. CH. E. Journal vol. 1, 513–521

(III B) 1976 Boissonade, J.: Aspect theoriques de la "double oscillation" dans les systemes dissipatifs chimiques J. Chimie de Physique, vol. 73, 540–544

(III B) 1980 Boissonade, J., De Kepper, P.: Transitions from Bistability to Limit Cycle Oscillations. Theoretical Analysis and Experimental Evidence in an Open Chemical System. J. Phys. Chem. vol. 84, 501–506

(III A) 1921 Bray, W. C.: A Periodic Reaction in Homogeneous Solution and its Relation to Catalysis. J. Amer. Chem. Soc. vol. 43, 1262–1267

(III A) 1931 Bray, W. C., Caulkins, A. L.: Reactions Involving Hydrogen Peroxide, Iodine and Iodate Ion: II. The Preparation of Iodic Acid. Preliminary Rate Measurements. J. Amer. Chem. Soc. vol. 53, 44–48

(III A) 1931 Bray, W. C., Liebhafsky, H. A.: Reactions Involving Hydrogen Peroxide, Iodine and Iodate Ion. I. Introduction, J. Amer. Chem. Soc. vol. 53, 38–44

(III C) 1935 Bray, W. C., Liebhafsky, H. A.: The Kinetic Salt Effect in the Fourth Order Reaction $BrO_3^- + Br^- + 2 H^+ \rightarrow$. Ionization Quotients for HSO_4^- at 25°, SO_4^- at 25°. J. Amer. Chem. Soc. vol. 57, 51–56

(III B) 1973 Briggs, T. S., Rauscher, W. C.: An Oscillating Iodine Clock, J. Chemical Education, vol. 50, 496

(III C) 1980 Burger, M., Koros, E.: Conditions for the Onset of Oscillations, J. Phys. Chem. vol. 84, 496–500

(III F) 1964 Chance, B., Hess, B., Betz, A.: DPNH Oscillations in a Cell-free Extract of S. Carlsbergensis, Biochem. and Biophys. Res. Commun. vol. 16, 182–187

(III A) 1967 Degn, H.: Evidence of a Branched Chain Reaction in the Oscillating Reaction of Hydrogen Peroxide, Iodine and Iodate, Acta Chem. Scand. vol. 21, 1057–1066

(III C) 1967 Degn, H.: Effect of Bromine Derivatives of Malonic Acid on the Oscillating Reaction of Malonic Acid, Cerium Ions and Bromate, Nature, vol. 213, 589–590

(III G) 1968 Degn, H.: Bistability Caused by Substrate Inhibition of Peroxidase in an Open Reaction System, Nature vol. 217, 1047–1050

(III G) 1969 Degn, H.: Compound III Kinetics and Chemiluminescence in Oscillatory Oxidation Reaction Catalyzed by Horseradish Peroxidase, Biochem. Biophysica Acta, vol. 180, 271–290

(III G) 1969 Degn, H., Mayer, D.: Theory of Oscillations in Peroxidase Catalyzed Oxidation Reactions in Open System, Biochimica et Biophysica Acta, vol. 180, 291–301

(R1) 1972 Degn, H.: Oscillating Chemical Reactions in Homogeneous Phase, J. Chem. Education, vol. 49, no. 5, 302–307

(III B) 1976 De Kepper, A. Pacault, Rossi, A.: Etude d'une reaction chimique periodique, multistationnaire et transitions, C. R. Acad. Sci. Paris, Ser. C, vol. 282, 199–204

(III H) 1974 DePoy, P. E., Mason, D. M.: Periodicity in Chemically Reacting Systems. In: Faraday Symposia of the Chemical Society, No. 9, Physical Chemistry of Oscillatory Penomena, 47–54

(III F) 1957 Duysens, L. N., Amesz, M. J.: Fluorecence Spectrophotometry of Reduced Phosphorpyridine Nucleotide in Intact Cells in the Near-ultraviolet and Violet Region., Biochem. Biophys. Acta, vol. 24, 19–26

(III F) 1973 Dynnik, V. V., Sel'kov, E. E., Semashko, L. R.: Analysis of the Adenine Nucleotide Effect on the Oscillatory Mechanism in Glycolysis, Studia Biophysica, vol. 41, 193–214

(III F) 1975-1 Dynnik, V. V., Sel'kov, E. E.: Generator of Oscillations in the Lower Part of the Glycolytic System, Biophysics vol. 20, 292–297 (Biofizika, 288–292)

(III F) 1975-2 Dynnik, V. V., Sel'kov, E. E.: Double-Frequency Oscillations in the Glycolytic System. Mathematical Model, Biophysics vol. 20, 297–302 (Biofizika, 293–297).

(III E) 1973 Eckert, E., Hlavacek, V., Marek, M.: Catalytic Oxidation of CO on $CuO \cdot Al_2O_3$, I. Reaction Rate Model Discrimination, Chem. Eng. Comm. vol. 1, 89–94. II. Measurement and Description of Hysteresis and Oscillations in a Laboratory Catalytic Recycle Reactors Chem. Eng. Com. vol. 1, 95–102

(III E) 1973 Eckert, E., Hlavacek, V., Kubicek, M., Sinkule, J.: Zur Kenntnis des Zweiphasenmodelis des Katalytischen Reaktors, Chem. Ing. Tech. vol. 45, 83–88

(III A) 1979 Edelson, D., Noyes, R. M.: Detailed Calculations Modeling the Oscillatory Bray-Liebhafsky Reaction, J. Phys. Chem. vol. 83, 212–220

(III E) 1976 Eigenberger, G.: Kinetic Instabilities in Catalytic Reactions-A Modeling Approach, 4th Int. Symp. on Chem. Reaction Engineering, Heidelberg, 290–299

(III C) 1972 Field, R. J., Koros, E., Noyes, R. M.: Oscillations in Chemical Systems II. Thorough Analysis of Temperal Oscillations in the Bromate-Cerium-Malonic Acid System., J. Amer. Chem. Soc. vol. 94, 8649–8664

(III C) 1974 Field, R. J., Noyes, R. M.: Oscillations in Chemical Systems, IV. Limit Cycle Behavior in a Model of Chemical Reaction, J. Chem. Phys. vol. 60, 1877–1884

(R 12) 1978 Franck, U. F.: Chemical Oscillations, Angew. Chem. vol. 90, 1–16. (International Edition in English vol. 17, 1–15)

(R 8) 1976 Goldbeter, A., Caplan, S. R.: Oscillatory Enzymes, Ann. Rev. of Biophys. and Bioeng. vol. 5, 449–476

(III F) 1964 Gosh, A., Chance, B.: Oscillations of Glycolytic Intermediates in Yeast Cells, Biochem. Biophys. Res. Commun. vol. 16, 174–181

(R 5) 1974 Gray, B. F., Aarons: L. J.: Small Parasitic Parameters and Chemical Oscillations. In: Physical Chemistry of Oscillatory Phenomena, Symposium of the Faraday Society, No. 9, Faraday Division, Chemical Society, London, 129–136

(R 15) 1980 Gray, B. F.: Thermokinetic Oscillations in Gaseous System Kinetics of Physico-chemicals Oscillations, Berichte der Bunsen-Gesellschaft für Physikalische Chemie, vol. 84, no. 4., 309–315

(III G) 1975 Gurel, Demet: Dynamics of Thyroglobulin Iodination, 5th Int. Biophysics Congress, Copenhagen, Abstract P-438, p. 124

(III G) 1976 Gurel, Demet, Gans, P. J.: Kinetics of Enzymatic Thyroid Iodination, NIH NIAMD 1F32 AM05373 (Unpublished work)

(III G) 1977 Gurel, Demet, McNelis, E.: Oscillating Reactions in the Iodination of Thyroglobulin, NIH NIAMD 1R01 AM20850 (Unpublished work)

(III D) 1965 Gurel, O., Lapidus, L.: Liapunov Stability Analysis of Systems with Limit Cycles, Chem. Eng. Symposium Seties, vol. 61, no. 55, 78–87

(R 2) 1972 Gurel, O.: Bifurcation Theory in Biochemical Dynamics, In: Analysis and Simulation of Biochemical Systems, (Hemker, H. C., Hess, B., eds.) FEBS vol. 25, North-Holland, Amsterdam, 81–85

(III J) 1973 Gurel, O.: Topological Dynamics in Neurobiology, Int. J. Neuroscience, vol. 6, 165–179

(R 6) 1975 Gurel, O.: Limit Cycles and Bifurcations in Biochemical Dynamics Biosystems, vol. 7, 83–91

(III J) 1977 Gurel, O.: Decomposed Partial Peeling and Limit Bundles Physics Letter, vol. 61A, 219–223

(R 16) 1979-1 Gurel, O.: Some New Types of Oscillations, Kinetics of Physico-chemical Oscillations Discussion Meeting held by Deutsche Bunsengesellschaft für Physikalische Chemie, Aachen, vol. II, 486–494

(VB) 1979-2 Gurel, O.: Poincare Bifurcation Analysis, In: Gurel and Rossler (Book B4) p. 5–22

(III J) 1979 Gurel, O., Rossler, O. E.: Bifurcation to Toroidal Surfaces, Math. Japonica, vol. 23, 491–507

(III J) 1981 Gurel, O.: Exploded Points, Z. Naturforschung, vol. 36A, 72–75

(III F) 1964 Higgins, J.: A Chemical Mechanism for Oscillations of Glycolytic Intermediates in Yeast Cells, Proc. N.A.S. (USA) vol. 51, 989–994

(III F) 1967 Higgins, J.: The Theory of Oscillating Reactions, J. Ind. Eng. Chem. vol. 59, no. 5, 18–62

(R 11) 1978 Hlavacek, V., Votruba, J.: Hysteresis and Periodic Activity Behavior in Catalytic Chemical Reaction Systems, Advances in Catalysis vol. 27, 59–96

(III E) 1972 Horak, J., Jiracek, F.: Dynamic Behavior of Catalytic Flow Reactors, Chem. React. Eng., Proc. 5th Europ. Symp., B8, 1–12

(III E) 1968 Hugo, P.: Dynamic Behavior of Strongly Exothermic Catalytic Reactions in Open Gas Circulations (In German), Chem. React. Eng., Proc. 4th Eur. Symp. Pergamon Press, Oxford, England, (1971) 459–472

(III E) 1970 Hugo, P.: Stabilität und Zeitverhalten von Durfenss-Kreislauf-Reaktoren, Ber. Binsengesellschaft, Phys. Chem. vol. 74, 121–127

(III E) 1972 Hugo, P., Jakubith, M.: Dynamisches Verhalten und Kinetik der Kohlenmonoxid-Oxidation am Platin-Katalysator, Chem.-Ing.-Tech. vol. 44, 383–387

(III F) 1975 Kaimachnikov, N. P., Sel'kov, E. E.: Hysteresis and Multiplicity of Dynamic States in an Open Two-Substrate Enzymatic Reaction with Substrate Depression, Biophysics, vol. 20, no. 4, 713–718 (Biofizika, 703–708)

(III E) 1979 Kaimachnikov, N. P., Schulmeister, T.: Evolution of the Limit Cycle in a Model of an Enzymatic Reaction with Substrate Deposition, Studia Biophysica, vol. 75, Heft 1, 41–50

(III K) 1978 Koros, E., Orban, M.: Uncatalyzed Oscillatory Chemical Reactions, Nature, vol. 273, 371–372

(III K) 1980 Koros, E., Orban, M., Habon, I.: Chemical Oscillations during the Uncatalyzed Reaction of Aromatic Compounds with Bromate, 3. Effect of One-Electron Redox Couples on Uncatalyzed Bromate Oscillators, J. Phys. Chem., vol. 84, 559–560

(III I) 1968 Lefever, R.: Stabilite de Structure dissipative, Acad. Royal des Science de Belgique, Class des Sciences, Bulletin, vol. 54, 712–719

(III I) 1971 Lefever, R., Nicolis, G.: Chemical Instabilities and Substained Oscillations. J. theor. biol. vol. 30, 267–284

(III A) 1931-1 Liebhafsky, H. A.: Reactions involving Hydrogen Peroxide, Iodine, and Iodate. III. The Reduction of Iodate ion by Hydrogen Peroxide, J. Amer. Chem. Soc. vol. 53, 896–911

(III A) 1931-2 Liebhafsky, H. A.: IV. The Oxidation of Iodine to Iodate by Hydrogen Peroxide, J. Amer. Chem. Soc. vol. 53, 2074–2090

(III A) 1967 Lindblad, P., Degn, H.: A Compiler for Digital Computation in Chemical Kinetics and its Application to Oscillatory Reaction Schemes, Acta Chem. Scand. vol. 21, 791–800

(III H) 1910-1 Lotka, A.: Contribution to the Theory of Periodic Reactions, J. Phys. Chem. vol. 14, 271–274

(III H) 1910-1 Lotka, A.: Contribution to the Theory of Periodic Reactions, J. Phys. Chem. vol. 14,

(III C) 1975 Marek, M., Svobodova, E.: Nonlinear Phenomena in Oscillatory Systems of Homogeneous Reactions-Experimental Observations, Biophysical Chemistry, vol. 3, 263–273

(III A) 1974 Matsuzaki, I., Nakajima, T., Liebhafsky, H. A.: The Mechanism of the Oscillatory Decomposition of Hydrogen Peroxide by the I_2-IO_3^--couple, Chem. Letters (Japan) 1463–1466

(III G) 1969 Nakamura, S., Yokota, K., Yamazaki, I.: Sustained Oscillations in a Lactoperoxidase, NADPH and O_2 system Nature vol. 222, 794

(R3) 1973 Nicolis, G., Portnow, J.: Chemical Oscillations, Chem. Reviews, vol. 73, no. 4, 365–384

(III C) 1972 Noyes, R. M., Field, R. J., Koros, E.: Oscillations in Chemical Systems, I. Detailed Mechanism in a System Showing Temporal Oscillations, J. Amer. Chem. Soc., vol. 94, 1394–1395

(R4) 1974 Noyes, R. M., Field, R. J.: Oscillatory Chemical Reactions, Ann. Rev. Phys. Chem. vol. 25, 95–119

(R9) 1977 Noyes, R. M.; Field, R. J.: Mechanism of Chemical Oscillations: Experimental Examples, Acc. Chem. Res. vol. 10, 273–280

(III G) 1977 Olsen, L. F., Degn, H.: Chaos in an Enzyme Reaction, Nature, vol. 267, 177–178

(III K) 1979 Orban, M., Koros, E.: Chemical Oscillations during the Uncatalyzed Reaction of Aromatic Compounds with Bromate, 1. Search for Chemical Oscillators, J. Phys. Chem. vol. 82, 1672–1674

(III B) 1975 Pacault, A., de Kepper, P., Hanusse, P.: Description d'un Systeme Chimique dissipatif. Illustration d'un temps thermodynamique, C. R. Acad. Sci. Paris Vol. 280B, 157–161

(III B) 1975 Pacault, A., de Kepper, P., Hanusse, P., Rossi, A.: Etude d'une reaction chimique periodic Diagramme des Etas, C. R. Acad. Sci. Paris vol. 281C, 215–220

(III E) 1977 Pikios, C. A., Luss, D.: Isothermal Concentration Oscillations on Catalytic Surfaces, Chem. Eng. Sci. vol. 32, 191–194

(V B) 1885 Poincare, H.: Sur l'equilibre d'une mass fluide animee d'un movement de rotation, Acta Mathematica, vol. 7, 259–380. (See also, Oeuvres, 1952, Gauthier-Villars, Paris, vol. 7, 40–140)

(III F) 1966 Pye, K., Chance, B.: Substained Sinusoidal Oscillation of Reduced Pyridine Nucleotide in a Cell-Free Extract of Saccharomyces Carlsbergensis, Proc. N.A.S. (USA), vol. 55, 888–894

(R14) 1977 Ray, H. W.: Bifurcation Phenomena in Critically Reacting Systems, In: Applications of Bifurcation Theory (P. H. Rabinowitz, ed.) Academic Press, 285–315

(IIIH) 1965 Rinker, R. G., Lynn, S., Mason, D. M., Corcoran, W. H.: Kinetics and Mechanism of the Thermal Decomposition of Sodium Dithionite in Aqueous Solution, Ind. Eng. Chem. Fundmtls., vol. 4, 282–288

(IIIJ) 1975 Rossler, O. E.: A Multivibrating Switching Network in Homogeneous Kinetics, Bull. Math. Biol., vol. 37, 181–192

(IIIJ) 1976-1 Rossler, O. E.: Chaotic Behavior in Simple Reaction Systems, Z. Naturforschung, vol. 31a, 259–264

(IIIJ) 1976-2 Rossler, O. E.: An Equation for Continuous Chaos Physics Letters, vol. 57a, 397–398

(IIIJ) 1976-3 Rossler, O. E.: Chemical Turbulence: Chaos in a Simple Reaction-Diffusion System, Z. Naturforschung, vol. 31a, 1168–1172

(IIIJ) 1976-4 Rossler, O. E.: Different Types of Chaos in Two Differential Equations, Z. Naturforschung, vol. 31a, p. 1664–1670

(IIIJ) 1977-1 Rossler, O. E.: Toroidal Oscillation in a 3-Variable Abstract Reaction System, Z. Naturforsch., vol. 32a, 299–301

(IIIJ) 1977-2, 3 Rossler, O. E.: Chaos in Abstract Kinetics: Two Prototypes, Bull. Math. Biol., vol. 39, 275–289

(IIIJ) 1977-4, 5 Rossler, O. E.: Continuous Chaos, In: Synergetics: A Workshop, (H, Haken, ed.) Noted in Physics, Springer-Verlag, p. 184–199

(IIIC) 1978 Rossler, O. E., Wegmann, K.: Chaos in the Zhabotinskii Reaction, Nature, vol. 271, 89–90

(III J) 1979-1 Rossler, O. E.: Continuous Chaos: Four Prototype Equations, In: Gurel and Rossler, (Book B4), 376–392

(R13) 1979-2 Rossler, O. E.: Chaos and Strange Attractors in Chemical Kinetics, Springer Series in Synergetics, vol. 3, 107–113

(IIID) 1948 Salnikov, I. E.: Thermodynamic Model of Homogeneous Periodic Reactions, (in Russian), Dokl. Akad. Nauk SSSR, vol. 60, 405–408

(IIID) 1949 Salnikov, I. E.: On the Theory of Periodic Course of Homogeneous Chemical Reactions, (in Russian), Zh. Fiz. Khim., vol. 23, 258–272

(R7) 1975 Schmitz, R. A.: Multiplicity, Stability and Sensitivity of States in Chemically Reacting Systems In: Chemical Reaction Engineering Reviews (ed. Hulburt, H. H.), Advances in Chemistry Series, 148. American Chemical Society, Washington, D.C., 156–211

(IIIF) 1978 Schulmeister, Th.: Chaos in a Lotka-Scheme with Depot, Studia Biophysica, vol. 72, 205–206

(IIIF) 1978 Schulmeister, Th., Sel'kov, E. E.: Folded Limit Cycles and Quasi-Stochastic Self-Oscillations in a Third Order Model of an Open Biochemical System, Studia Biophysica, vol. 72, 111–112

(IIIF) 1968-1 Sel'kov, E. E.: Self-Oscillations in Glycolysis. A Simple Kinetic Model, Eur. J. Biochem., vol. 4, 79–86

(IIIF) 1968-2 Sel'kov, E. E.: Self-Oscillations in Glycolysis. Simple Single-Frequency Model, Molecular Biology, vol. 2, 208–221 (Molekulyarnaya Biologiya 252–266)

(IIIF) 1972 Sel'kov, E. E.: Nonlinearity of Multienzyme Systems In: Analysis and Simulation of Biochemical Systems, (Hemker, H. C., Hess, B., eds.) FEBS 25, North-Holland, Amsterdam, 145–161

(IIIF) 1973 Sel'kov, E. E., Betz, A.: On the Mechanism of Single-Frequency Glycolytic Oscillations, In: Biological and Biochemical Oscillations, (Chance, et al. eds.) Academic Press, 197–220

(IIIF) 1979 Sel'kov, E. E., Dynnik, S. N., Kirsta, Y. B.: Qualitative Investigation of a Mathematical Model of the Open Futile Cycle of fructose-6-P → fructose-1,6-P_2, Biophysics, vol. 24, 443–450 (Biofizika, 431–437)

(IIIF) 1980 Sel'kov, E. E.: Instability and Self-Oscillations in the Cell Energy Metabolism, Berichte der Bunsen-Gesellschaft für Physikalische Chemie, vol. 84, no. 4, 399–406

(III A) 1976 Sharma, K. R., Noyes, R. M.: Oscillations in Chemical Systems, 13. A Detailed Molecular Mechanism for the Bray-Liebhafsky Reaction of Iodate and Hydrogen Peroxide, J. Amer. Chem. Soc. vol. 98, 4345–4361

(R10) 1977 Sheintuch, M., Schmitz, R. A.: Oscillations in Catalytic Reactions, Catalysis Reviews, vol. 15, 107–172

(III A) 1911 Skrabal, A.: Zur Kenntnis der unterhalogenigen Säuren und der Hypohalogenite, V. Die Kinetik der Jodatbildung aus Iod und Hydroxylion, Monat. Chemie, vol. 32. 815–903

(III C) 1915 Skrabal, A., Weberitsch, S. R.: Zur Kenntnis der Halogensauerstoffverbindungen, IX. Die Kinetik der Bromat-Bromidreaktion Monat. Chemie vol. 36, 211–235. X. Die Kinetik der Bromatbildung aus Brom Monat. Chemie vol. 36, 237–256

(V2) 1980 Tockstein, A., Komers, K.: Biomolecular Kinetic Scheme with a Stable and an Unstable Limit Cycle of Two Oscillatory Components Collection Czechoslovak Chem-Commun. vol. 45, 2135–2142

(III D) 1974 Uppal, A, Ray, W. H., Poore, A. B.: On the Dynamic Behavior of Continuous Stirred Tank Reactors, Chem. Eng. Sci. vol. 29, 967–985

(III D) 1976 Uppal, A, Ray, W. H., Poore, A. B.: The Classification of the Dynamic Behavior of Continuous Stirred Tank Reactors-Influence of Reactor Residue Time Chem. Eng. Sci. vol. 31, 205–214

(III C) 1973 Vavilin, V. A., Zhabotinskii, A. M., Zaikin, A. N.: A Study of a Self-Oscillatory Chemical Reaction, I. The Autonomous System, In: Biological and Biochemical Oscillations (Ed. B. Chance), 71–79

(III C) 1978 Wegmann, K., Rossler, O. E.: Different Kinds of Chaotic Oscillations in the Belousov-Zhabotinskii Reaction, Z. Naturforschung A, vol. 33A, no. 10, 1170–1183

(III J) 1980 Willamowski, Rossler, O. E.: Irregular Oscillations in a Realistic Abstract Quadratic Mass Action System, Z. Naturforsch., vol. 35a, 317–318

(III G) 1965 Yamazaki, L., Yokoya, K., Nakajima, R.: Oscillatory Oxidations of Reduced Pyridine Nucleotide by Peroxidase, Biochim. Biophys. Res. Commun., vol. 21, 582–586

(III G) 1967 Yamazaki, I., Yokota, K.: Analysis of the Conditions Causing the Oscillatory Oxidation of Reduced Nicotinamide-Adenine Dinucleotide by Biochem. Biophys. Acta, vol. 132, 310–320

(III E) 1974 Yang, C. H.: On the Explosion, Glow and Oscillation Phenomena in the Oxidation of Carbon Monoxide, Combustion and Flame, vol. 23, 97–108

(III C) 1932 Young, H. A., Bray, W. C.: The Rate of the Fourth Order Reaction between Bromic and Hydrobromic Acid. The Kinetic Salt Effect. J. Amer. Chem. Soc., vol. 54, 4284–4296

(III C) 1973 Zaikin, A. N., Zhabotinskii, A. M.: A Study of a Self-oscillatory Chemical Reaction, In: B. Chance, et al. (Book-B1) 81–88

(III C) 1964-1 Zhabotinskii, A. M.: Periodic Course of Oxidation of Malonic Acid in Solution, (An Investigation of the Kinetics of the Reaction of Belousov), Biophysics, vol. 9, 329–335, (Biofizika, 9, 306–311)

(III C) 1964-2 Zhabotinskii, A. M.: Periodic Oxidation Reaction in Liquid Phase, (in Russian) Dokl. Akad. Nauk SSSR vol. 157, no. 2, 392–395

Books

(B1) 1973 Chance, B., Ghosh, A. K., Pye, E. K., Hess, B. (Eds.): Biological and Biochemical Oscillators, Academic Press, New York
Academic Press, New York

(B2) 1974 Physical Chemistry of Oscillatory Phenomena, Symposia of the Faraday Society, no. 9, The Faraday Division, Chemical Society, London

(B3) 1979 Pacault, A., Vidal, C. (Eds.): Synergetics, Far from Equilibrium, Springer-Verlag Series in Synergetics, vol. 3, Springer, Berlin, Heidelberg, New York

(B4) 1979 Gurel, O., Rossler, O. E. (Eds.): Bifurcation Theory and Applications in Scientific Disciplines N.Y. Acad. of Sciences Annals no. 316

Recent Developments in Chemical Oscillations

Demet Gurel,[1] and Okan Gurel[2]

1 Department of Chemistry, New York University, New York, N.Y. 10003, USA
2 Cambridge Scientific Center, IBM Corporation, Cambridge, Mass 02142, USA

Table of Contents

I Introduction

The goal of the previous article in this volume was to introduce the field of oscillatory chemical reactions to the reader in an organized format. The original and subsequent significant contributions only were included by sifting and selecting through a rather large literature.

In this article an attempt is made to include some new contributions to bring the field of oscillations in the dynamics of chemical reactions up to date. To provide a continuation, the basic structure of the paper follows the outline of the previous article which will be referred to as (G & G). Moreover, some recent studies have also been added to the field, and they are included as additional sections. These new sections are: L. Oxidation by Chlorite; M. Miscellaneous Studies; and N. General Models and Mathematical Techniques.

II Review Articles

Since the preparation of the article (G & G), there have been numerous review articles on some of the reactions, appearing in the literature. In Table II (Table numbers coincide with the Section numbers) a list of these review articles and the reactions discussed are given. As seen in this table, the topic of oscillations has been expanding such that these reviews. although limited in scope, attest to the world wide interest in oscillatory chemical reactions.

Table II. Review Articles on Oscillatory Chemical Reactions

Review by		Reaction[a]												Number of References	Language
		A	B	C	D	E	F	G	H	I	J	K	L		
Boiteux et al.	80						×							66	English
Brisset	80						×	×						14	French
Klonowski	80						×							19	Polish
Noyes	80	×	×	×			×		×	×		×		84	English
Ruoff	81	×		×										31	Norwegian
Seno, Iwamoto	81			×										22	Japanese
Slinko, Slinko	80				×									69	Russian (Eng. trs)
Taranenko	80						×	×						39	Russian
Tsuda	80			×				×			×			40	Japanese
Zhabotinskii	80			×										68	English
Zhabotinskii	82			×										54	Hungarian

a The letters indicating the reactions are as in Section III: A. Iodate Catalyzed Decomposition of Hydrogen Peroxide (Bray-Liebhavsky Reaction), B. An Oscillating Iodine Clock (Briggs-Rauscher Reaction), C. Oxidation of Malonic Acid by Bromate (Belousov-Zhabotinskii Reaction). D. Continuous Stirred Tank Reactor, E. Solid Catalyzed Reactions, F. Oscillations in Glycolysis, G. Peroxidase Catalyzed Reactions, H. Decomposition of Sodium Dithionite, I. Bimolecular Model, J. Abstract Reaction Systems, K. Uncatalyzed Reactions, L. Oxidation by Chlorite.

III Reactions and Models Exhibiting Oscillations

The reactions where oscillations are either observed or predicted have been increasingly attracting researchers into the field. Most of the new studies have been directed toward understanding the mechanism behind the oscillatory phenomena. There are experimental as well as theoretical investigations evaluating the effects of changes in reaction variables and parameters on the oscillatory behavior of the systems. In this section some of the recent contributions to the types of reactions and models outlined in (G & G) are briefly discussed, and some newly proposed reactions are also added to the list. Wherever it is possible, these contributions are grouped under subsections.

A Iodate Catalyzed Decomposition of Hydrogen Peroxide (Bray-Liebhavsky Reaction)

Although this reaction is the oldest among the oscillatory reactions known to chemists, it has not yet been fully explored. However, a complete understanding of this oscillatory systems might possibly shed light on iodide-peroxide interactions in biological systems. We group the recent studies into two subsections.

A.1 Studies on the Reaction Mechanism

Reinvestigating the original pioneering work on the reaction exhibiting oscillations, Liebhafsky, et al. (1981) discussed a skeleton mechanism explaining basically the nonoscillatory behavior of the reaction, "smooth catalysis" of hydrogen peroxide decomposition. However, by computer analysis of this model, they found that following the smooth catalysis, after a long interval, oscillatory catalysis appears.

Petrenko, et al. (1982) proposed a nine step mechanism involving four intermediates (HIO_2, I^-, HOI, I_2) for the reaction of IO_3^- with H_2O_2 in acid solution. By determining the kinetic parameters for this mechanism, oscillations for the four intermediates were obtained and reported.

A.2 Some Variations and their Effects

Cooke (1980-1) investigated the effect of Cu(II) and chloride ions on the iodate-peroxide reaction in the presence and absence of Mn(II). Cooke (1980-2) further studied hydrogen peroxide-iodic acid-Mn(II)-organic species system with respect to

1. The Bray-Liebhavsky reaction ($H_2O_2 - H^+ - IO_3^-$) and
2. The Belousov-Zhabotinskii reaction, $Ce(III) - BrO_3^- - H^+ - CH_2(CO_2H)_2$.

The behavior of the redox potential oscillations under the effects of hydrogen peroxide, iodate ions, Mn(II) ions, acetone, and sulfuric acid was examined. The interrelationship between redox potential oscillations and iodine oscillations was studied. Both the chloride ion and Cu(II) inhibit the iodate-hydrogen peroxide reaction where Mn(II) catalyzed production of iodine is the key step.

Veljkovic-Slobodanka (1981) observed the oscillatory reaction in mixtures of potassium iodate and hydrogen peroxide in the presence of enhanced oxygen circulation at higher temperatures. By controlling the concentration changes of I^- potentiometrically an explicit acceleration of oscillations was determined.

Zueva and Protopopov studied the anion effect on the redox reaction of $KIO_3/H_2O_2/Cys$ HCl in H_2SO_4 (1982-1), and observed increases in amplitude of oscillations. Also, the effect of cysteine hydrochloride on the alterations of oscillations in the same system were studied, (1982-2).

By adding hydrogen peroxide to a moderately acidic solution of potassium iodate and perchloric acid changes in behavior observed at 50 °C with an iodide ion specific electrode, were reported by Odutola et al. (1982). As the concentrations of H^+ and H_2O_2 were increased, following the addition of hydrogen peroxide, oscillations in the potential with varying periods were recorded.

B An Oscillating Iodine Clock (Briggs-Rauscher Reaction)

Studies on this reaction are grouped under three subsections. In B.1, work directed towards the elucidation of the reaction mechanism are discussed. In B.2, those experiments with different substrates and media are listed. In Table III.B.2 we summarize some of the variations in the substrate and the medium in which the reaction has been observed. In addition, an experimental technique is included in B.3.

Table III.B.2. New Substrates and Media for the Briggs-Rauscher Reaction

Substrate	Medium	Reference
Acetoacetic ester	Phosphoric acid or Sulfuric acid Hydrochloric acid or Phosphoric acid/Cl^-	Dutt & Banerjee (1981-1) Dutt & Banerjee (1981-2)
Acetone Acetylacetone Ethylacetoacetate		Dutt & Banerjee (1982)
Methylmalonic acid		Furrow (1981)
Crotonic acid		Furrow (1982)
Phenol Acrylamide Oxalic acid		Furrow & Noyes (1982-2)

B.1 Studies on the Reaction Mechanism

Furrow and Noyes (1982-1) considered the iodate-peroxide subsystem (Liebhavsky, 1931) as

(A) $2IO_3^- + 5 H_2O_2 + 2 H^+ \rightarrow I_2 + 5 O_2 + 6 H_2O$

coupled with

(B) $2 H_2O_2 \rightarrow 2 H_2O + O_2$

in which oscillations had been observed earlier by Bray (1921). A mechanism of Mn(II) catalysis of reaction (A) above, as well as uncatalyzed reaction (A) were discussed at length. Mn(II) ion catalyzed iodate oxidation of peroxide is almost 10^3 times faster than the uncatalyzed reaction.

Noyes and Furrow (1982) modelled the essential mechanistic features of the full iodate-peroxide-Mn(II)-malonic acid $(IO_3^- - H_2O_2 - Mn(II) - CH_2(CO_2H)_2)$ system with a mechanism involving 30 pseudo-elementary processes. Eleven of these processes are believed to generate the oscillatory behavior, observed experimentally. The rate constants of seven of these processes were determined experimentally. The other four values are assigned to create a system that mimics the essential features of oscillations. This mechanism differs from those of other known oscillators in that both radical and nonradical paths generate the same net chemical change.

De Kepper and Epstein (1982) identified ten reactions in the Briggs-Rauscher (B-R) system. These reactions overlap those proposed by Noyes and Furrow (1982) and by Cooke (1980-2). The regions of oscillations for the initial values of I_2 and IO_3^- were identified. By numerical simulations oscillations were obtained.

Due to the intermediate I_2, the B-R reaction is very photosensitive. Hosrthemke (1980), in a continuously stirred tank reactor (CSTR) measured the optical density of the system against the mean incident light intensity and showed that the white noise (environmental) effect on chemical reactions must be included in a realistic of such reactions.

B.2 Experiments with Different Substrates and Media

Acetoacetic ester, $MeCOCH_2CO_2Et$, was found to be a new substrate for the B-R reaction in both phosphoric, and sulfuric acid media by Dutt and Banerjee (1981-1). They determined the limits of the concentration range of each component. The mechanism was related to the model given by Noyes, et al. for the Belousov-Zhabotinskii reaction. Dutt and Banerjee (1981-2) were also able to observe oscillations in hydrochloric acid medium and in phosphoric acid that included chloride ions, Cl^-/H_3PO_4. The induction period for oscillations increases with increased chloride ion concentration and oscillations are not observed when the chloride concentration reaches a level above 0.04 M, [See also D. O. Cooke, React. Kinet. Catal. Lett. 3 (1975) 377].

Dutt and Banerjee (1982) studied temperature dependence of the B-R reaction with substrates such as, malonic acid, acetone, acetylacetone [Acetylacetone was first used in Dutt and Banerjee (1980)] and ethylacetoacetate within 280–320 K range. With increased temperature faster oscillations were observed.

Furrow (1981) observed that in the B-R reaction with methylmalonic acid, the oscillatory periods were lengthened by increasing IO_3^- and H^+ concentrations and shortened by increasing H_2O_2, methylmalonic acid and Mn(II) concentrations.

Furrow (1982) studied a subsystem of the B-R reaction, without malonic acid, where iodine production was effected by introducing crotonic acid (trans-2-butenoic acid) to the system. Although oscillations are not observed, this study may prove to be useful in understanding the mechanism of oscillating systems.

Furrow and Noyes (1982-2) studied the effect of addition of various organic and inorganic reagents in removing iodine present as I_2 and HOI. For example, reagents very effective in removing HOI were found to have very definite influence on the iodate-peroxide-Mn(II) subsystem where as removal of iodide and iodine had almost none. The authors thus concluded that HOI is an important intermediate in Mn(II) catalysis of reaction (A). These reagents are malonic acid, $CH_2(COOH)_2$, methylmalonic acid, $CH_3CH(COOH)_2$, crotonic acid, trans-$CH_3CH=CHCO_2H$, phenol, C_6H_5OH, acrylamide, $CH_2=CHCONH_2$, oxalic acid $(CO_2H)_2$, pyrophosphate, $P_2O_7^{4-}$, silver ion, Ag^+, dichromate, $Cr_2O_7^{2-}$, and chloride ion.

B.3 Experimental Techniques

Betteridge et al. (1981) using the technique of acoustic emission, studied the B-R reaction and reported the acoustic energy trace of the system. The trace follows the oscillations of the solution as the color changes between blue and colorless states.

C Oxidation of Malonic Acid by Bromate (Belousov-Zhabotinskii Reaction)

Since the discovery of the Belousov-Zhabotinskii (B-Z) reaction a large number of variations of this reaction has been investigated by numerous researchers. Both experimental and theoretical investigations outnumber those for any other oscillatory reaction. The reason for this extensive interest comes from the fact that the B-Z reaction is very rich in its interesting dynamical behavior and its variations are quite extensive. In this section research done during the period of 1980-82 is discussed briefly.

At the start of this three year period, Noyes (1980) presented a generalized mechanism explaining the oscillations observed in the B-Z reactions. The thermodynamic and kinetic constraints and the limitations of the mechanism were discussed in detail. The reactions discussed are:
1) Original B-Z reaction where metal ion catalyzed organic substrate is brominated by enolization and bromide is liberated when the organic bromate reacts with the oxidized form of the catalyst, [See G & G].
2) B-Z reaction with phenols and anilines as substrate, [Orban, Koros, Noyes, J. Phys. Chem. 82 (1978) 1672].
3) B-Z reaction with mixed substrates: Tartaric acid/acetone [Rastogi, R. P., Singh, H. J. and Singh, A. K., Kinetics of Physicochemical Oscillations, Preprints of Submitted Papers. Aachen Discussion Meeting of Deutsche Bunsengesellschaft für Physicalische Chemie, (1979) 98–107], oxalic acid/acetone [Nos#ticzius, Mag. Kem. Foly 85 (1979) 330].
4) Ce(III) catalyzed B-Z reaction with oxalic acid as substrate, [Nos#ticzius and Bodiss, J. Am. Chem. Soc. 101 (1979) 3177].
5) B-Z reaction with added silver nitrate exhibits suppressed oscillations of a bromide-specific electrode while the potential of the platinum electrode still oscillates, [Nos#ticzius, J. Am. Chem. Soc. 101 (1979) 3660].

The above reactions 1–4 switch rapidly between conditions of relatively large and small concentrations of bromide ion whenever $[Br^-]$ exceeds a critical value. However in reaction 5, switching is controlled by HOBr rather than Br^-. A detailed generalized mechanism was given to explain reactions 1–4 by modifying the models of the first and the second reactions.

The general mechanism consists of four component stochiometric processes as (A) through (D). Denoting HOBr, $HBrO_2$, $BrO_2\cdot$ and BrO_3^- as Br(I), Br(III), Br(IV) and Br(V), respectively, the four processes are

$$Br(V) + 2\,Br^- \rightarrow 3\,Br(I) \tag{A}$$

$$Br(V) + 4\,Red \rightarrow Br(I) + 4\,Ox \tag{B}$$

$$Br(I) + n\,Ox \rightarrow Br^- + X \tag{C}$$

$$p\,Br(I) + q\,Br^- \rightarrow inert\ Br \tag{D}$$

Red is the reduced form of catalyst metal ion. In type 2, Red is the organic substrate itself. Here X corresponds to a mixture of compounds.

In section C we summarize the recent contributions under subsections. The general considerations on the mechanism of the B-Z reaction are in four subsections. Following these general considerations, we have sections: C.2. Mathematical models and techniques, C.3. Experiments with different substrates, C.4. Experiments with different catalysts, and C.5. Horatian oscillations in bromate oxidations. The name *horatian* has recently been proposed for replacing the term *chaotic*. Here we simply state that throughout this article instead of chaotic oscillations, the term *horatian* oscillations will be used.

C.1 General Considerations on Mechanisms and Study Techniques

Under the general considerations we group various studies in separate subsections. Some researchers investigated the role of stirring in this reaction. Coupling of such reactions has also been studied. In another subsection those studies on the bromide ion and its role in the reaction are grouped. The influence of other factors such as diffusion, heat exchange, iodide ion addition is discussed in another subsection. General experimental techniques are grouped in the last subsection.

C.1.1 Studies with Stirring and Coupling

Orban et al. (1982-2) discovered that in a CSTR within an extremely narrow range of flow rates and input concentrations a system containing BrO_3^-, Br^- and Mn(II) or Ce(III) exhibits oscillations in the potential of either a Pt redox or Br^- selective electrode. Existence of oscillations was predicted by the model calculations of Bar-Eli [Bar-Eli in Vidal and Pacault (1981) 228–239]. The bromate oscillators such as the B-Z reaction were derived from this fundamental system by adding feedback species which enlarges the region of critical space in which oscillations occur.

Geiseler (1982-2) studied the autocatalytic oxidation of Mn(II) by acidic bromate in the presence of bromide inhibitor and observed oscillations in Br^- concentrations.

Referring to an earlier study by Marek and Stuchl [Biophys. Chem. 3 (1975) 241–248], who reported the oscillating concentration changes of Ce(III)/Ce(IV) ions in continuously stirred tank reactors, Nakajima and Sawada (1980) studied the interaction of two coupled oscillating systems of the B-Z reactions using H_2SO_4, $Ce_2(SO_4)_3$, $KBrO_3$, malonic acid and ferroin solution as the chemical reagents in two coupled CSTRs. They obtained experimentally a mapping diagram relating the ratio w_2/w_1 (the ratio of the natural frequencies w_2 and w_1 of the two reactors before coupling) and the size of the window area S_{1-2} between the two coupled systems representing the strength of the coupling. Corresponding to various frequency ratios they measured and reported varying types of voltage differences between the two Pt electrodes. In certain regions of the mapping diagrams a horatian behavior of the differences was observed and related analogously to Tomita and Kai's [Tomita K, and Kai T, Phys. Lett. A 66 (1978) 91–93, and Progress Theor. Phys. 61 (1979) 54–73] region where a horatian response is predicted for the coupled system, when one reactor is externally excited.

Bar-Eli and Geiseler (1981), examined two coupled CSTRs containing Ce(III) and BrO_3^- and Br^- in sulfuric acid medium for oscillations and transitions from one steady state to another. The CSTR's were coupled in two different configurations, in parallel and in series. No oscillations were reported under tested constraints and configurations. Geiseler (1982-1) discovered and presented the initial investigation of limit cycle oscillations in the stirred flow oxidation of Ce(III) by acidic BrO_3^-. The region in the parametric planes of the inflow concentrations of BrO_3^- and Br^- versus malonic acid, corresponding to oscillatory behavior of the system were also investigated, and oscillations in Ce(IV) concentrations were observed.

C.1.2 Studies on Bromate/Bromide Ions

Investigating the role of bromomalonic acid during the induction period of the B-Z reaction Burger and Koros (1980-1, 2) found that bromomalonic acid concentration has to reach a certain level for oscillations to start. This "crucial" concentration in turn is affected by changes in the acid concentration of the medium.

Observing that the feed-back role played by bromide ion in the B-Z reaction, contrary to the previous schemes, may not explain recent experimental findings, Nosticzius and Bodiss (1980) proposed the Lotka-Volterra model for a B-Z reaction with combined oxalic acid/acetone substrate. The reaction scheme becomes, for $X = HBrO_2$, $Y = HOBr$,

$$[A] + X \rightarrow 2X,$$
$$X + Y \rightarrow 2Y, \qquad [A] = 3\,H^+ + BrO_3^-$$
$$Y \rightarrow [B]. \qquad [B] = CH_2BrCOCH_3 + H_2O$$

Although this scheme is oscillatory, it does not possess a limit cycle (periodic) solution. Therefore, this was later extended to a four-variable Lotka-Volterra scheme as

$$[A] + X \rightarrow W \rightarrow 2X, \qquad [D] = CH_3COCH_3$$
$$X + Y \rightarrow Z \rightarrow 2Y, \qquad W = Br_2O_4$$
$$[D] + Y \rightarrow [B]. \qquad Z = Br_2O_2$$

which possesses a periodic solution. Working with the bromide selective electrode, Noszticzius (1981) showed that the response of the electrode potential is probably due to hypobromous acid and no information about the bromide concentration could be obtained under the given scheme. Thus it was concluded that the bromide ion is not a controlling intermediate in the B-Z reaction. The applicability of the

Lotka-Volterra model to the B-Z reaction with other substrates was verified by Nosticzius and Feller (1982).

Ganapathisubramanian and Noyes (1982-1, 2), examing the experimental environment, confirmed that the results obtained in measuring oscillating bromide ion concentrations are reliable and mechanistic interpretations are valid.

Zhabotinskii, et al. (1982), following the concentration of Ce(IV) in a CSTR with open up inflows of BrO_3^-, Ce(III) and Br$^-$ by spectrophotometry, with and without added bromomalonic acid, found that production of Br$^-$ is not proportional to bromomalonic acid concentrations. A radical mechanism for bromate reduction was also proposed.

Kovalenko et al. (1980) reported the relation between malonic acid and BrO_3^- concentrations for the oscillatory regions of the system $BrO_3^- - CH_2(COOH)_2 - Mn(II)$ (or Ferroin). In addition, the induction period, the period and the amplitude of the oscillations of concentrations of Mn(III), were determined as a function of a) malonic acid for various BrO_3^- concentrations and b) BrO_3^- for various malonic acid concentrations.

C.1.3 Factors Influencing Oscillations

Oscillations in the potential of the B-Z reaction are found to be influenced by the addition of iodide ion to the system. By increasing the amount of iodide ion in the system Kőrős and Varga (1982) recorded these variations and reported that high frequency oscillations precede the expected oscillations of the B-Z reaction. Similar high frequency oscillations were also observed by addition of monoiodomalonic acid (IMA) to the B-Z system.

Vidal and Noyau (1980) investigated the qualitative and quantitative influence of heat exchange in the B-Z reaction in a CSTR reactor and concluded that the B-Z oscillations are intrinsic and insensitive to heat exchange.

Jorne (1980) confirmed the findings of earlier researchers that the trigger wave propagation in the ferroin catalyzed B-Z reaction is caused by the coupling between autocatalytic mechanism and diffusion.

Ganapathisubramanian and Noyes (1982-3) studied the B-Z reaction to discover that the reaction has complexities which are hard to explain, however the interpretation of the basic mechanism is not affected by these features.

C.1.4 Experimental Techniques

Adamcikova and Treindl (1976) recorded the time dependence of the limiting diffusion current of the Mn(II)/Mn(III) catalyst ions by a polarograph with a rotating platinum electrode. The results of the kinetic measurements were also discussed by referring to the B-Z model, [See G & G].

Kovalenko and Tikonova (1980) used inert electrodes and discovered a correlation between the oscillations in the redox potential and the concentration changes in the B-Z reactants. The graphite and vitreous carbon electrodes were used to measure the concentration ratio of the redox forms of the catalyst metal ions while the concentration of bromate reduction products were measured by the inert platinum electrodes.

Botre et al. (1981), using the B-Z reaction as the model reaction, studied energy changes in the system under different experimental conditions by measuring electrical potentials.

Keszthelyi et al. (1981) applied x-ray microanalysis technique to the study of a B-Z-type reaction of (HNO$_3$/KBrO$_3$/[Fe(phe)$_3^{2+}$]/malonic acid system). However such a technique requiring a dyring process may result in interpretations based on somewhat altered structures.

Schlueter and Weiss (1981), in studying the oxidation of malonic acid with BrO$_3^-$ in the presence of Mn(II)/Mn(III) as catalyst applied the nuclear magnetic relaxation titration (NMRT) to follow the oscillations in the B-Z system. Oscillations of the signal amplitude and those in relaxation rate of the ^{1}H-nuclear magnetization in aqueous solution were measured and reported.

C.2 Mathematical Models and Techniques

In addition to extensive experimental studies on the B-Z reaction several mathematical models have been discussed either as part of the experimental results or as general studies reflecting the results available in the literature. Some of the studies are discussed in this section.

Reducing the four-dimensional scheme of Nosticzius (1981) to a three dimensional one, differential equations for the B-Z reaction were given by Noszticzius and Farkas (1981) as

$$dx/dt = a + bx - xy$$
$$dy/dt = cx + 2dx - xy - ey$$
$$dz/dt = xy - dz$$

which produces a limit cycle oscillation and is based on a scheme where bromide ion does not play a role.

In terms of $X = $ [HBrO$_2$], $Y = $ [Br$^-$], and $Z = $ [2 Ce(IV)], De Kepper and Boissonade (1981) proposed a three dimensional model with six parameters depending on the reaction rates, and compared computational results to the experimental observations in a CSTR environment favorably.

Geiseler and Bar-Eli (1981) devised an improved form of the earlier models. Application of this model to the BrO$_3^-$—Ce(III)-malonic acid-H$_2$SO$_4$ system in a CSTR was illustrated. The region of oscillations in the parameter planes of the inflow concentrations of BrO$_3^-$ and Br$^-$ versus malonic acid were investigated. Oscillations in Ce(IV) concentrations were observed.

Based on two concentrations as X_k and Y_k of the kth system, Nakajima and Sawada (1981) proposed a mathematical model of a coupled system ($k = 1, 2$) and interpreted their earlier experimental work, see Section C.1.1.

In an attempt to explain the horatian oscillations due to the B-Z reaction in a well-stirred continuous flow reactor reported by R. A. Schmitz et al., Iwamoto and Seno (1981) proposed a reaction model and a two dimensional mathematical model.

Working with the model of B-Z reaction, Sakanoue and Endo (1982) showed by computer simulation the coexistence of a stable and an unstable limit cycle. The existence of an unstable oscillating object between two stable objects had been demonstrated ten years earlier by computer simulation for the Hodgkin-Huxley equations, [see Gurel, Int. J. Neurosc. 5 (1973) 281].

Edelson (1981), using numerical analysis techniques studied the relationship between the period of the B-Z oscillations and the rate constants or initial conditions of the reaction. The principal rate-controlling steps were confirmed numerically to be the enolization and bromination of malonic acid.

C.3 Experiments with Different Substrates

An extensive set of studies is devoted to the variations on the reaction by replacing malonic acid as the substrate by another organic (or inorganic) substrate. These investigations are grouped in this subsection and summarized in Table III.C.3.

Table III.C.3. Belousov-Zhabotinskii Reaction and Different Substrates

Organic Substrate	Inorg. Subs.	Catalyst	References
Malonic acid ⎫ Citric acid ⎬ Maleic acid ⎭		Ce(III, IV)	Zhabotinskii (Original)
Malic acid ⎫ Bromomalic acid ⎬ Dibromomalic acid ⎭			[Kasperek & Bruice (1971)]
Acetylenedicarboxylic acid			[Beck & Varady Mag Chem Foly 81 (1975) 519]
Acyclic and cyclic ketones			[Stroot & Janjic Helv Chim Acta 58 (1976) 116]
Phenol, aniline and derivatives			[Orban & Kőrős J. Phys. Chem. 82 (1978) 1672]
Tartaric acid/Acetone			[Rastogi et al. Aachen Meeting (1979) 98-107]
Oxalic acid or Glyoxalic acid			[Nosticzius & Bodiss J. Am. Chem. Soc. 101 (1979) 3177]
Oxalacetic acid		Mn(III, II)	Maselko (1980-3)
Oxalic acid	⎧ ⎨ ⎩	Ce(III) or Mn(II)	Sevcik & Adamcikova (1982)
Tartaric acid		Mn(II) ⎫ ⎬	Adamcikova & Sevcik (1982-1)
Oxalic acid/Acetone		⎭	Nosticzius & Bodis (1980)
Cyclohexanone ⎫ Cyclopentanone ⎬		Ce(IV)	Farage & Janic (1980-1, 2)
Acetylacetone		Mn(II, III)	Rastogi & Rastogi (1980)
Phosphonoacetic acid		Ce(IV)	Habashi-Krayenbuhl & Janjic (1982)
α-Ketoglutaric acid		Ce(III, IV)	Treindl & Dorovky (1981, 1982)
Salicylic acid ⎫ 5-Sulfosalicylic acid ⎬		Ferroin	Gupta & Srinivasulu (1982)
2,4-Pentanedione			⎧ Treindle & Fabian (1980) ⎨ ⎩ Treindle & Kaplan (1981)
Pyrogallol		Mn(II)	Habon & Kőrős (1979)
	NaH_2PO_2 (Hypophosphate)	Mn(II)	Adamcikova & Sevcik (1982-2)

Farage and Janjic (1980-1) observed oscillations in redox potential for the B-Z reaction where bromate oxidation of cerium in cyclohexanone or cyclopentanone in acidic solution (H_2SO_4) was studied, and later Farage and Janjic (1980-2) compared the alterations in oscillations of these two systems under different concentrations of reactants. Furthermore, Farage and Janjic (1981) reported that for the BrO_3^- — Ce(IV)—CH_2CO_2H (B-Z reaction) system, oxygen inhibited oscillations and the mechanical stirring in a nitrogen atmosphere had no significant effect. However for the BrO_3^-—Ce(IV)-cyclohexanone or cyclopentanone systems under specific experimental conditions, stirring was found to play a determining part, and oxygen to have no notable effect on the oscillations. Patonay and Noszticzius (1981) confirmed the observations of Farage and Janjic (1980) that stirring considerably reduces the number of oscillations of the B-Z reaction.

Replacing malonic acid by phosphonoacetic acid (PAA), Habashi-Krayenbuhl and Janjic (1982) observed oscillations in redox potential. For various concentrations of the reactants (PAA, Ce(IV), BrO_3^- and H_2SO_4), the induction period, period of oscillations and redox potential and the number of oscillations were studied.

Rastogi and Rastogi (1980) investigated the oscillatory reaction in acetylacetone/$KBrO_3$/Mn(III)/H_2SO_4 system and reported oscillations in Br^- and Mn(III)/Mn(II) and rate of temperature rise.

Sevcik and Adamcikova (1982) investigated the catalyzed oxidation of oxalic acid with BrO_3^- under constant nitrogen flow. Changes in concentration of catalyst, Br_2 and BrO_3^- were followed by the polarographic method. Either Ce(III) or Mn(II) could be used as catalyst in the reaction. Oscillatory behavior of oxalic acid and malonic acid under heterogeneous conditions was compared. In a more recent study, Adamcikova and Sevcik (1982-1) reported that oscillations were observed in the presence of Mn(II) ions when tartaric acid was used as the substrate. The evidence suggests that the oscillation cycle consists of the following reactions:

$$Mn(II) \xrightarrow{BrO_3^-} Mn(III) + Br_2 \tag{A}$$

$$Mn(III) \xrightarrow[\text{tartaric acid}]{H^+} Mn(II) \tag{B}$$

It is also reported that the reduced form of this catalyst, Mn(II), is regenerated by reaction (B) to reenter into reaction (A). Since in the case of Ce(III) regeneration of the reduced form is a slow reaction, Ce(III) does not produce the same result. In addition, Adamcikova and Sevcik (1982-2) found that an inorganic compound, hypophosphate, NaH_2PO_2 when substituted for the organic reducing material generates oscillations in a B-Z type reaction. Periodic Br_2 evolution was recorded polarographically.

Treindl and Fabian (1980) studied the effect of oxygen on parameters of the B-Z reaction in the presence of Ce(IV)/Ce(III) redox catalyst and malonic acid, citric acid or 2,4-pentanedione as substrates. They concluded that the effect of oxygen was in its catalytic influence on the oxidation of the substrate with Ce(IV) ion. Under this influence, the number of oscillations as well as the induction period and the first oscillation period diminished.

A B-Z type reaction involving α-ketoglutaric acid as a substrate was studied by Treindl and Dorovky (1981, 1982). The polarographic method with a rotating platinum electrode was used in following the temporal oscillations both of Ce(IV) and Br_2 concentrations simultaneously.

Treindl and Kaplan (1981) studied the kinetics of oxidation of 2,4-pentanedione with Ce(IV) ions. The modified B-Z reaction with 2,4-pentanedione as substrate exhibited oscillations with an increasing amplitude even in the absence of stirring.

The role of bromide ion in the B-Z reaction was questioned by Noszticzius and Bodiss (1980) while working with a B-Z reaction where combined substrate oxalic acid/acetone was used as the organic substrate.

Habon and Kőrős (1979) found that pyrogallol could be a substrate in a Mn(II) catalyzed B-Z reaction. The critical Br^- concentrations, the effect of inhibitors such as oxygen and light were studied.

Zueva and Sipershtein (1980) studied oscillations in bromate/Ce(IV)/citric acid system. For various concentration ratios between potassium bromate, citric acid and Ce(IV) they measured variations in amplitudes of bromide oscillations and that of reagent concentrations. The influence of additions of sec-butanol on *starting* oscillations in the B-Z system with oxalic acid, succinic acid and tartaric acid, and *altering* the oscillations in systems with malonic acid and citric acid was studied by Zueva and Sipershtein (1981). This influence was shown to vary depending on concentration ratios between the *sec*-butanol and the substrate of the B-Z reaction being studied. Furthermore, the variations depend on the order in which the substrate, oxidizing agent and additional agents are introduced into the reaction.

C.4 Experiments with Different Catalysts

While there are numerous studies on B-Z reactions with different substrates, there are as many investigations of reactions with different catalysts. A summary of these studies is given in Table III.C.4.

Bolletta and Balzani (1982) observed oscillating chemiluminescence in a B-Z reaction with tris (2,2'-bipyridine ruthenium (II), $Ru(bpy)_3^{2+}$, [First used by Demas and Diemente J. Chem. Ed. 50 (1973) 357]. This is the first example of oscillating chemiluminescence in a B-Z reaction.

D'Alba and Serravalle (1981) compared the effect of various catalysts on the B-Z reaction by using an electrochemical method, [see Botre et al. (1981)]. The catalysts compared are $Ce(SO_4)_2$, ferroin and $FeSO_4$ and their various combinations. The number of oscillations, the average frequencies and the duration of oscillations were measured and tabulated for different cases.

Gupta and Srinivasulu (1981), in addition to uncatalyzed B-Z type reaction (see Section K), studied the reaction with ferroin as catalyst.

Yoshida and Ushiki (1982) compared the kinetics of the B-Z reaction with Ce(III) or $[Fe(phen)_3]^{2+}$ as catalyst, and found substantial differences in kinetic constants. Similarly, Yoshikawa (1982) studied the effects of temperature on the frequencies of oscillations in the B-Z reaction. They reported that for a reaction where Ce(III) was used as the catalyst, the activation energy range was fixed. However, when $Fe(phen)_3^{2+}$ or $Fe(bpy)_3^{2+}$ was used as catalyst, the apparent activation energy depended on the concentration of the catalyst.

Table III.C.4. Belousov-Zhabotinskii Reaction with Different Catalysts

Organic Substrate	Catalyst	References
Malonic acid	$[Ag(bipy)_2]^+/[Ag(bipy)_2]^{2+}$	Kuhnert & Pehl (1981-1)
	$[Os(bipy)_3] SO_4 \cdot 8 H_2O$	Kuhnert & Pehl (1981-2)
	Cr-bipyridine	
	$Ce(SO_4)_2$	
	Ferroin	D'Alba & Serravella (1981)
	$FeSO_4$ and combinations	
		Yoshida & Ushiki (1982)
	$[Fe(phen)_3]^{2+}$	[Rovinskii & Zhabotinskii Theor. Expr. Chem. 15 (1978) 17]
	$[Fe(phen)_3]^{2+}$ and $[Fe(bpy)_3]^{2+}$	Yoshikawa (1982)
		Tikhonova & Zayats (1980)
	$Ru(bpy)_3^{2+}$	Bolletta & Balzani (1982)
		[Demas & Diemente (1973)]
	0-phenanthroline, 2,2-dipyridyl, p-phenylenediamine, 8-hydroxyquinoline, Aniline complexes of Co(II), Ni(II), Cu(II)	Handlirova & Tockstein (1980)
	Bis-bipyridine-silver	Kuhnert & Pehl (1981-1)
	Bipyridine complexes of osmium and chromium	Kuhnert & Pehl (1981-2)
	Tetraazamacrocyclic complexes of Ni(II) (Nitrogen is needed)	Yatsimirskii et al. (1981-2)
	Combined Ce(III), Ce(IV) and Ferroin	Kovalenko et al. (1981)
	Combined Ce(III), Ce(IV) and Ferrein	Yatsimirskii et al. (1981-1)
	Combined Ce(III), Ce(IV) and Mn(II), Mn(III)	Tikhonova et al. (1981)
Malic acid	Combined Ce(III), Ce(IV) and Mn(II), Mn(III)	Ramaswamy et al. (1980)

Tikhonova and Zayats (1980) experimented with $Ru(bpy)_3^{2+}$/KBrO$_3$/malonic acid system and compared the oscillations with those of the system with Ce$_2$(SO$_4$)$_3$.

Bis-bipyridine-silver complexes were found to catalyze the B-Z (with malonic acid as substrate) reaction by Kuhnert and Pehl (1981-1). The reaction was shown to proceed in a heterogeneous medium due to the insolubility of the silver complexes. When organic compounds such as citric acid and 2,4-pentanedione, ethylacetoacetate and racemic malic acid were used as substrates, the oscillatory behavior was not observed. Kuhnert and Pehl (1981-2) also observed that the bipyridine complexes of chromium and osmium catalyze the B-Z reaction.

Handlirova and Tockstein (1980) experimented with o-phenanthroline, 2,2-dipyridyl, p-phenylenediamine, 8-hydroxyquinoline and aniline complexes of Co(II), Ni(II), and Cu(II). They observed that some of these complexes showed catalytic activity.

Experimenting with the combined catalytic action of Ce(III, IV) and ferroin on the oscillating reaction system bromate/malonic acid/H$_2$SO$_4$ for different ratios of ferroin to Ce(III) concentrations, Kovalenko et al. (1981) reported different types of oscillations and also indicated the intervals of the ratios where similarities in oscillations were observed.

Tikhonova et al. (1981) studied the combined catalytic action of Ce(III, IV) and Mn(II, III) in oscillating redox reaction in the B-Z system by measuring Br⁻ potential. As catalysts, both metal ions act independently. Similar to Tikhonova et al. (1981), Ramaswamy et al. (1980) examined the system of combined Mn(III)/Mn(II) and Ce(IV)/Ce(III) with malic acid as substrate, and followed the oscillating potentials of a platinum electrode.

Yatsimirskii, et al. (1981-1) studied the B-Z reaction under different combinations of ferroin/ferriin and Ce(III)—Ce(IV) couples used as catalyst. The changes in the amplitudes of the oscillating bromide concentrations were recorded.

In the experiments by Yatsimirskii et al. (1981-2), the tetraaza macrocyclic complexes of Ni(II) were found to be catalysts in the BrO_3^-/malonic acid/H_2SO_4 system. A stream of nitrogen was necessary for oscillations to appear.

C.5 Horatian Oscillations in Oxidations by Bromate

Maselko (1980-1) studied the B-Z system in a CSTR and measured Pt and bromide electrode potentials. Depending on the reagent concentration in the inlet of the reactor and the retention time, several types of limit cycles were detected in the Mn(III)—Br⁻ plane. The multiple peak oscillations result in the formation of loops in the limit cycle, Fig. III.C.1. The conceptual ideas on the horation oscillations were also discussed. Maselko also used citric acid (1980-2), oxalic acid (1980-3), and malic acid (1982) as substrates. The bifurcation diagrams on the parameter spaces, concentration of citric (or malic) acid versus that of $KBrO_3$, were plotted. Regions of various oscillating and nonoscillating behaviors were identified.

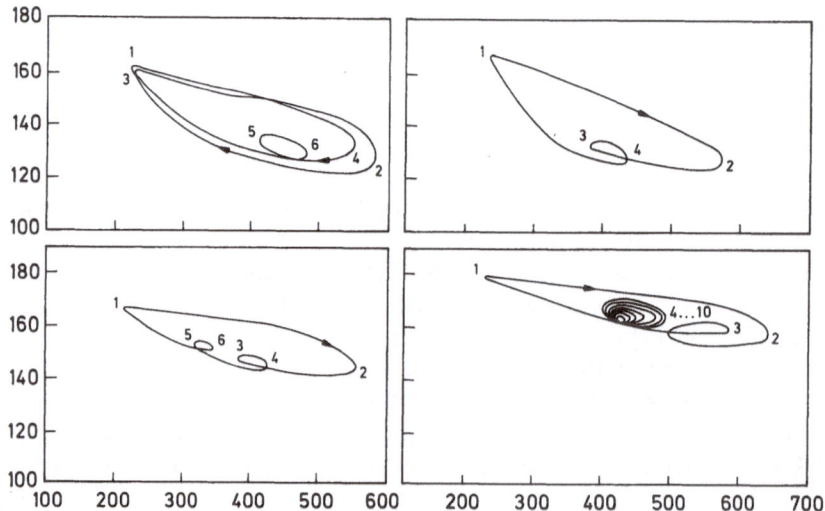

Fig. III.C.1. The concentrations of Br⁻ versus Mn(III) as multiple peak oscillations with looped limit cycles. (After Maselko (1980-1))

Experimenting with the B-Z reaction in a CSTR at various residence times for the same inlet concentrations and temperature, Hudson and Mankin (1981) obtained and reported horatian oscillations. The measurements of the bromide ion electrode potential and the platinum electrode potential were recorded. Calculating the time derivative of Pt, or the time delay, Pt(t-10), three dimensional phase space plots were presented. With the time delay as the third variable a horatian solution was obtained, Fig. III.C.2.

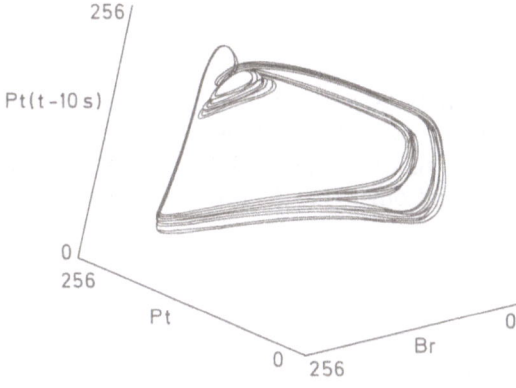

Fig. III.C.2. A horatian solution of the Belousov-Zhabotinskii Reaction, Pt, Br, and Pt(t-10) as variables. (After Hudson & Mankin (1981))

Ganapathisubramanian and Noyes (1982-4) constructed a model consisting of seven differential equations. The numerical solutions of these equations did not correlate with the horatian behavior observed by Hudson-Mankin (1981) in B-Z reaction in a flow reactor.

Iwamoto and Seno (1981) model, (see Section C.2) was designed to explain the horatian oscillations due to the B-Z reaction in a CSTR.

Nakajima and Sawada (1980) (See Sect. C.1.1, and C.2) experimented with two coupled CSTRs, and for a certain region of the mapping diagram they obtained horatian oscillations.

Starting with the B-Z model, Fujisaka and Yamada (1980) studied the horatian behavior of coupled B-Z systems.

Working with a B-Z reaction, catalyzed by combined Ce(III) and Fe(II) in a CSTR, Nagashima (1980) observed different oscillatory behaviors of the reaction. For certain concentrations of Fe(II), they found alternating periodic and horatian oscillations. For low concentration regions multipeak oscillations, and for high concentration regions only periodic oscillations were observed. Later, Nagashima (1982) observed horatian oscillations under external periodic perturbations.

Experimenting with the B-Z reaction in a CSTR Turner et al. (1981-1) observed a sequence of alternating periodic and horatian oscillations as the residence time was increased. The same sequence was predicted by a numerical study of a four-variable model describing the primary reaction steps. Such alternating phenomenon was studied also by Pikovskii (1981).

Vidal et al. (1980, 1981) experimented with the cerium bromate/malonic acid system in a CSTR and by increasing the flow rate detected bifurcation of a one-frequency regular oscillatory pattern to a quasi periodic regime with two independent frequencies. However, they could not verify the sustained quasi-periodic oscillations. Later, Vidal et al. (1982) observed periodic and horatian oscillations, alternating regularly when the flow through the reactor was increased.

Pomeau et al. (1981) discussed the intermittent transition to horatian behavior in a B-Z reaction. Ce(IV) oscillations were recorded for different residence times and essential experimental results were shown to be consistent with the intermittency type transition to horatian oscillations.

D Continuous Stirred Tank Reactor

Although the oscillatory behavior of continuously stirred tank reactor had been discussed in chemical engineering long before the current studies on homogeneous chemical systems, [see G & G], it is recently that the role of stirring and of a CSTR environment in reactions such as the B-Z and others has been recognized and used extensively in experiments.

D.1 Belousov-Zhabotinskii Reaction in a CSTR

In studying the horatian behavior of the B-Z reaction Maselko (1980-1) experimented in a CSTR environment (see Sect. C.5). Later, Maselko (1980-2) and (1982) studied the bifurcations by distinguishing the regions experimentally. Similarly, experiments by Vidal et al. (1980, 1981 and 1982) in studying the horatian behavior of the reaction were all run in a CSTR, see Section C.5.

Orban et al. (1982-2) (see Sect. C.1.1) showed that in a CSTR within an extremely narrow range of flow rates and input concentrations, a B-Z system with Mn(II) or Ce(III) as catalysts exhibits oscillations in the potential of either Pt redox or Br⁻ selective electrodes.

Bar-Eli and Geiseler (1981), Geiseler and Bar-Eli (1981), Geiseler (1982-1, 2), De Kepper and Boissonade (1981), and Hudson and Mankin (1981) conducted their B-Z reactions in a CSTR. Iwamoto and Seno (1981) model was also based on a B-Z reaction in a CSTR.

Nakajima and Sawada (1980) experimented with two coupled CSTRs.

D.2 Oxidation by Chlorite in a CSTR

These reactions, were all run in a CSTR where CSTR environment is necessary for the oscillatory behavior, see Section L.

D.3 Other Experiments in a CSTR

Crooke et al. (1980) discussed the fermentation process, the growth of yeast, in a continuously stirred tank fed with glucose, minerals and vitamins. Defining X as the concentration of the cells and S as that of the substrate (nutrient) a model consisting of two nonlinear differential equations of the first order was given and sustained oscillations in X and S were obtained. Specifically the cases of one limit cycle and two limit cycles were illustrated.

Vayenas et al. (1980) studied experimentally the oxidation of ethylene on poly-crystal Pt film in a CSTR reactor, and Vayenas et al. (1981) developed a model to explain the oscillatory phenomena. Lignola, et al. (1980) reported oscillations during propane oxidation in a CSTR.

Wirges (1980) investigated the oscillatory behavior of the catalytic decomposition of hydrogen peroxide by $Fe(NO_3)_3 \cdot 9 H_2O$ in a nitric acid solution in a CSTR. The critical value of the reaction rate that results in oscillations of the temperature and the conversion was identified. The comparison of the experimental values with the computer results showed considerable deviations for the temperature amplitude. An extended mathematical model taking into account the evaporation of water was proposed giving a better agreement between theory and experiment. Starting with a dimensionless form of the material and energy balance equations, Hugo and Wirges (1980) studied conversion and temperature differences numerically. They determined the oscillatory behavior of the temperature amplitudes.

Morton and Goodman (1981-1, 2) studied the CO oxidation in a CSTR, see Sections E.1 and 10.

Kahlert et al. (1981) proposed a simplified model for nonisothermic CSTR reaction exhibiting horatian oscillations.

E Solid Catalyzed Reactions

The study of solid catalyzed (heterogeneous) reactions and their oscillatory behavior has been developing independently without much interaction with the extensive studies in homogeneous reactions such as the B-Z system. In this section we subgroup these studies by the substrate reacting in an oscillatory manner. These reactions are summarized in Table III.E.

E.1 Carbon Monoxide Oxidation

Bond et al. (1982) discussed the oxidation of CO theoretically and experimentally. They observed the limit cycle behavior as found by Yang's model [Yang, C. H. Combust. Flame 23 (1974) 97] and conducted experiments recording light intensity of the oscillating glow of dry CO and O_2 mixture.

Bouillon et al. (1982) examined the oscillating CO oxidation on Pt as a possible candidate for the model by Takoudis, et al. (1981), see Section E.10. The results of Monte Carlo simulation of this model reaction were compared with their numerical results.

Oscillatory behavior of catalytic CO oxidation on Pt surfaces was found by Ertl et al. (1982) to be influenced by the reverse phase transition of the Pt surface.

Keil and Wiecke (1980) investigated oscillations in the CO oxidation on Pt catalysts in a tubular reactor under isothermal conditions. At lower and higher CO contents the kinetics of the reaction could be described uniformly by a Langmuir-Hinshelwood relationship.

Liao and Wolf (1982) experimented with CO oxidation on Pt/γ-Al_2O_3 catalyst. The effects of temperature and of O_2/CO ratios on the oscillations were observed. Under these effects, different observations for fresh, pretreated and poisoned catalysts were reported.

To study the oxidation of CO, Morton and Goodman (1981-2) set up an experimental system by using CO, oxygen and butene in a CSTR containing a Pt catalyst.

Table III.E. Solid Catalyzed Reactions

Reaction	Catalyst	References
CO Oxidation		
Periodic Oscillations:	Pt	Bond et al. (1982)
	Pt	Bouillon et al. (1982)
	Pt	Keil & Wicke (1980)
	Pt/γ-Al$_2$O$_3$	Liao & Wolf (1982)
	Pt	Morton & Goodman (1981-2)
	Pt	Zhang (1980)
	Pt, Pd	Sales et al. (1981)
	Pd	Rajagopalan (1981)
	Pd, Ir	Turner et al. (1981-2)
Horation Oscillations:	Pt	Rathousky et al. (1980)
	Pt/Al$_2$O$_3$	Rathousky & Hlavacek (1981, 1982)
	Pd/α-Al$_2$O$_3$	Hlavacek & Rathousky (1982)
		Jensen & Ray (1980-1)
	Pd	Rajagopalan (1981)
H Oxidation		
Periodic Oscillations:	Pt	Kasemo et al. (1980)
	Pt	Wicke et al. (1980)
	Pt, Pd	Rajagopalan (1981)
	Pd	Rajagopalan et al. (1980)
Horation Oscillations:	Ni	Kurtanjek, et al. (1980-1, 2)
	Ni	Sault & Masel (1982)
	Ni	Schmitz et al. (1980)
	[Me]	Chumakov et al. (1980)
		Jensen & Ray (1980-1, 2)
NO Reduction		
Periodic Oscillations:	Pt	Hegedus et al. (1980)
Horatian Oscillations:	Pt	Adlhoch, et al. (1981)
Methanol Oxidation	Pd	Jaeger et al. (1981)
Ethylene Oxidation	Pt	Vayenas et al. (1980)
Ethylene Hydrogenation	Ni—Al$_2$O$_3$	Niiyama & Suzuki (1982)
Propylene Oxidation (Horatian)	Pt	Sheintuch & Luus (1981)
Butane Oxidation		
Periodic Oscillations:		Caprio et al. (1981)
Horatian Oscillations:		Jensen & Ray (1980-1, 2)
Cyclohexane Oxidation		
Periodic Oscillations:	KY Zeolite	Ukharskii et al. (1981)
Horatian Oscillations:		Jensen & Ray (1980-1, 2)

The ranges of temperature and butene feed corresponding to different flow rates, were experimentally recorded and regions of oscillations were determined. For a fixed temperature, (150 °C) and 1% butene feed, corresponding to increasing flow rates different oscillatory behaviors in each of the CO, O_2, CO_2 concentrations were observed. A model consisting of four nonlinear equations was formulated and corresponding to various combinations of parameters, oscillations in a limit cycle form were obtained. The model was used to detect oscillations obtained for this experiment as well as some observations reported earlier in the literature.

Zhang (1980) studied the kinetics of CO oxidation on a Pt ribbon, and found that in this system gas impurities were responsible for oscillations.

Sales et al. (1981) tested the hypothesis of oscillations in the rate of CO_2 production being in part due to the alternate oxidation and reduction at a Pt or Pd surface layer.

Turner et al. (1981-2) observed oscillations in the rate of CO_2 production over Pd and Ir catalysts, similar to earlier observations of oscillations with Pt as catalyst. By changing the gas temperature T_g, variations in oscillations of CO oxidation measured by changes in catalyst temperature, T_c were observed.

Rajagopalan (1981) investigated the oxidation of CO over a Pd wire and H_2 oxidation over Pt and Pd wires. Both periodic oscillations with one or multiple peaks for each cycle and horatian oscillations were observed. The observed periods for CO oxidation were short (in seconds) while those of hydrogen oxidation varied from minutes to an hour. A mathematical model was given closely corresponding to the experimental results.

Jensen and Ray (1980), Rajagopalan (1981), Rathousky et al. (1980), Rathousky and Hlavacek (1981, 1982), Hlavacek and Rathousky (1982) studied CO oxidation and observed horatian oscillations, see Section E.11.

E.2 Hydrogen Oxidation

Kasemo et al. (1980) described an experimental system to study catalytic reactions. Gas samples close to the catalyst were examined. The oxidation of H_2 on polycrystalline Pt was studied by this method and oscillations in the concentration of H_2 were observed.

Rajagopalan et al. (1980) referring to Kurtanjek (1980)'s work observed sustained oscillations of the reaction rate during the oxidation of hydrogen on a palladium wire. The temperature of the catalyst was maintained constant. It was concluded that the reaction rate oscillations observed during H_2 oxidation over metal catalysts were associated with a transition between oxidized and reduced catalyst surface states as previously suggested by Kurtanjek (1980), see Section E.11. Rajagopalan (1981) studied H_2 oxidation, see Section E.1.

Wicke et al. (1980) presented a study of oscillations during oxidation of hydrogen on Pt catalyst. Oscillations of catalyst temperature at different O_2 concentrations were shown to differ from each other. Following a discussion of the mechanism of the oscillations, they also compared the H_2 oxidation experiments with CO oxidation.

Kurtanjek (1980-1), Sault and Masel (1982), Schmitz et al. (1980), Chumakov et al. (1980), Jensen and Ray (1980-1, 2) studied H_2 oxidation and observed horatian oscillations, see Section E.11.

E.3 Nitrogen Oxide Reduction by CO

Hegedus et al. (1980) carried out experiments over a Pt-alumina catalyst exposed to mixtures of NO, CO, and O_2 by periodically switching the stoichiometry of the feedstream between reducing and oxidizing conditions. The concentrations of species on the catalyst surface were measured and found to be oscillating. The results were viewed as proof that the catalyst surface events determine the transient response characteristics of the system.

Adlhoch et al. (1981) studied horatian oscillations, see Section E.11.

E.4 Methanol Oxidation

Oxidation of methanol to CO_2 and H_2O on a supported Pd as catalyst exhibits oscillations. Jaeger et al. (1981) measured the temperature of the catalyst corresponding to various frequencies and reported periodic and complex oscillations in temperature.

E.5 Ethylene Oxidation

Vayenas, et al. (1980) studied experimentally and proposed a model (1981) to explain the oscillatory behavior of ethylene oxidation on polycrystalline Pt, in a CSTR. Chang and Aluko (1982) argued that these results were incorrect, however Vayenas et al. (1982) refuted this contradiction on theoretical and physical grounds.

E.6 Ethylene Hydrogenation

Niiyama and Suzuki (1982) studied hydrogenation of ethylene over $Ni-Al_2O_3$. Oscillations in particle temperature were observed. Similarly in a packed bed flow reactor, a multiple-particle system, the temperature of the catalyst bed and the conversion of ethylene, both showed oscillations.

E.7 Propylene Oxidation

Sheintuch and Luus (1981) studied propylene oxidation and observed horatian oscillations, see Section E.11.

E.8 Butane Oxidation

Dependence of isobutane oxidation on temperature and the residence time in a CSTR was discussed by Caprio et al. (1981). Different oscillations were detected.

Jensen and Ray (1980-1, 2) studied horatian oscillations, see Section E.11

E.9 Cyclohexane Oxidation

Ukharskii et al. (1981) reported oscillations in CO_2 concentrations during oxidation of cyclohexane on KY zeolite.

Jensen and Ray (1980-1, 2) reported horatian oscillations in this reaction, see Section E. 11.

E.10 Theoretical Models

Discussing the models based on Langmuir-Hinshelwood mechanisms involving an adsorbed species which acts as a buffer, Lynch and Wanke (1981) showed the difference between the predictions by general models and their simplified forms given earlier. They indicated that the simplifications result in significantly altered predictions.

To study a class of mechanisms for isothermal heterogeneous catalysis in a CSTR, Morton and Goodman (1981-1) analyzed the stability and bifurcation of simple models. The limit cycle solutions of the governing mass balance equations were shown to exist. An elementary step model with the stoichiometry of CO oxidation was shown to exhibit oscillations at suitable parameter values. By computer simulation limit cycles were obtained.

Developing methods for studying the bifurcation behavior of tubular reactors, Jensen and Ray (1982-2) discussed these in detail and reported periodic oscillating solutions resulting from bifurcations.

Takoudis et al. (1981) proposed a model for a bimolecular Langmuir-Hinshelwood surface reaction with two empty sites in its reaction step. The two chemisorbed species were assumed to adsorb competitively on the surface. The two dimensional model with reaction rates as parameters were shown to exhibit oscillations. Bifurcation of this model was also discussed. Takoudis et al. (1982) described a procedure for obtaining necessary and sufficient conditions for the existence of periodic solutions in surface reactions with constant temperature. An analytic method for the analysis of bifurcation to periodic solutions was developed.

Sales et al. (1982) presented a simple physical model to explain oscillatory oxidation of carbon monoxide over Pt, Pd, and Ir catalysts. The model is based on a kinetic model incorporating a Langmuir-Hinshelwood reaction mechanism and the alternate oxidation and reduction of the catalyst. Simulation results of these three coupled differential equations (of oxidation of CO) model are shown to fit experimental observations.

Suhl (1981) proposed a model consisting of three differential equations incorporating two different adsorbtion sites for oxygen.

Sheintuch (1981) analyzed an oscillatory kinetic mechanism with surface oxide and gas phase reactant as variables, and showed that depending on operating conditions an asymmetric state of surface oxide is reached. The asymmetric state was found to be stable, except near the bifurcation point where it might be oscillatory. A mathematical model was developed. Sheintuch and Pismen (1981) investigated the existence of inhomogeneous surface states for three oscillatory kinetic models, i.e. autocatalytic gas-phase variable, autocatalytic surface variable and two surface variables. Sheintuch (1982) also analyzed an oscillatory kinetic mechanism by employing two surface concentrations as variables and the mechanism was simulated by the proposed model and discussed.

Rajagopalan (1981) presented a mathematical model for CO and H_2 oxidations, see Section E.1.

Zioudas (1980) developed a model of three differential equations and obtained both periodic and horatian oscillations, see Section E. 11.

Related to studies on horatian oscillations, Chumakov et al. (1980) and Schmitz et al. (1980) proposed mathematical models for hydrogen oxidation, see Section E.11.

E.11 Horatian Oscillations in Solid Catalyzed Reactions

Rathousky, et al. (1980) studying CO oxidation in three different experimental setups (a recycle reactor packed by Pt, a tubular packed bed reactor and honeycomb matrix) observed horatian behavior for high inlet temperatures.

Rathousky and Hlavacek (1981) presented two mathematical models to illustrate the fact that the influence of adsorbed species on the rate of an isothermal catalytic reaction may lead to a complex dynamic pattern including multiplicity of steady states and oscillatory states. Multiple oscillations and horatian behavior can not be calculated from the models. Rathousky and Hlavacek (1982) studied CO oxidation on Pt/Al$_2$O$_3$ catalyst and observed changes in oscillations due to the variations in inlet temperature. For a narrow range they observed horatian behavior. Experiments show that interaction of two oscillatory processes cause horatian behavior.

Hlavacek and Rathousky (1982) studied CO oxidation on an α-Al$_2$O$_3$ honeycomb matrix impregnated by Pd. By measuring catalyst temperature and outlet conversion over a range of inlet CO concentrations and temperatures they observed oscillations including horatian behavior in exit conversion. For higher inlet temperatures the observed horatian oscillations became more "symmetrical" and finally disappeared.

Kurtanjek, et al. (1980-1 and 2) observed horatian oscillations in constant potential difference (CPD) and O$_2$ concentrations in oxidation of hydrogen on nickel foil in a CSTR. They suggested that the oscillations are induced by cyclic oxidation and reduction of the surface.

Chumakov et al. (1980) studied the reaction of oxidation of H$_2$ on a metal catalyst [Me]. The reaction scheme was given as:

$$
\begin{aligned}
&\text{H}_2 \quad + a[\text{Me}] \rightleftarrows 2[\text{HMe}]\,, \\
&\text{O}_2 \quad + 2[\text{Me}] \rightleftarrows 2[\text{OMe}]\,, \\
&2[\text{HMe}] + [\text{OMe}] \rightarrow 3[\text{Me}] \quad + \text{H}_2\text{O}\,, \\
&[\text{OMe}] \;+ \text{H}_2 \quad \rightarrow [\text{Me}] \quad + \text{H}_2\text{O}\,, \\
&[\text{Me}_n] \quad + [\text{OMe}] \rightleftarrows [\text{Me}_n\text{O}] \;+ [\text{Me}]
\end{aligned}
$$

(A)

(B)

where [Me$_n$] is near surface layer of the catalyst. The mathematical model corresponding to the reaction scheme is:

$$
\begin{aligned}
\mathrm{d}x/\mathrm{d}t &= k_1(1 - x - y)^2 - k_{-1}x^2 - 2k_3x^2y \\
\mathrm{d}y/\mathrm{d}t &= k_2(1 - x - y)^2 - k_4y - k_3x^2y \\
\mathrm{d}z/\mathrm{d}t &= e(y(1 - z) - \alpha z(1 - x - y))\,.
\end{aligned}
$$

They concluded that the existence of horatian oscillations indicates that the kinetic model of the catalytic system contains equations of higher than second order.

Sault and Masel (1982) measured horatian oscillations in temperature for the oxidation of hydrogen in nickel foil. The effect of surface pretreatment on ignition instabilities leading to self-sustained oscillations were examined.

Adlhoch, et al. (1981) discussed the reduction of nitric oxide by CO on a polycrystal platinum ribbon in an open system. It was concluded that the oscillations (even horatian) resulted from the interaction of adsorbed gases and the surface structure.

Sheintuch and Luus (1981), as part of the search for the oxidation of hydrocarbons, studied the isothermal oxidation of propylene over platinum wire and observed horatian oscillations.

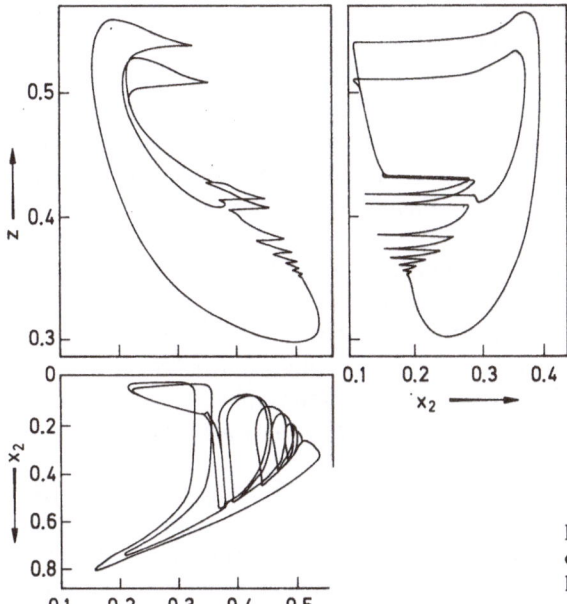

Fig. III.E.11. The Complex limit cycle of Chumakov et al. model [After Chumakov et al. (1980)]

Jensen and Ray (1980-1) studied the oxidation of butane, cyclohexane, hydrogen and carbonmonoxide, and reported horatian oscillations. A new model predicting these oscillations was proposed, (1980-2), and it was further studied in (1982-1).

Rajagopalan (1981) observed horatian oscillations in CO and H_2 oxidations, see Section E.1.

Schmitz et al. (1980) studied the nickel catalyzed oxidation of hydrogen. Oscillatory behavior including multipeak periodic and horatian, were observed.

In studying the dynamic behavior of gas/solid catalyzed reactions, Zioudas (1981) proposed a model consisting of three ordinary differential equations, and obtained multiple peak and horatian oscillations. The horatian oscillations resulted at the end of bifurcations from single peak to 2, 3, 4 and 8 peak periodic oscillations.

F Oscillations in Glycolysis

Termonia and Ross (1981-1, 2) developed a reaction scheme coupling phosphofructo-kinase and pyruvate kinase reactions by referring to experimental observations of known activations and inhibitions of enzymes by metabolites. By numerical analysis of the rate equations they confirmed the oscillations in the concentrations of fructose-6-phosphate, pyruvate, phosphoenolpyruvate, fructose 1,6-biphosphate, and ADP.

Termonia and Ross (1981-3) working with Higgins' mechanism of the phospho-fructokinase reaction of the glycolytic pathway where by imposing a periodic supply of fructose-6-phosphate entrainment is realized, constructed a mathematical model and computationally obtained an appropriate solution.

In a set of studies, the efficiency of dissipating chemical energy was demonstrated. Richter and Ross (1980) by a heat engine analogy formulated phosphofructokinase reaction converting fructose-6-phosphate to fructose 1,6-diphosphate as the glycolysis machine. Richter et al. (1981) starting with Sel'kov's model of glycolysis (see G & G) concluded that the system was efficiencient in converting the chemical energy. Again in Richter and Ross (1981) it was suggested that the mechanism generating the oscillations in glycolysis may have evolved so that the dissipation of free energy is reduced.

An analysis of the influence of enzyme cooperativity in the mathematical model of product activated oscillatory glycolysis reaction was made by Goldbetter and Venieratos (1980). They established the relationship between the instabilities and the value of the Hill coefficient in the allosteric model for phosphofructokinase.

An open two-substrate reaction catalyzed by phosphofructokinase of E. coli was modelled by Malkova et al. (1980). The parameter regions where oscillations are observed were also determined.

As a model to explain oscillations in an enzyme reaction Chay (1981) proposed a model based on a feedback mechanism of proton gradients across a membrane. The activity of the key enzyme in the reaction depends on the pH of the inner compartment. Oscillations predicted in pH as well as enzyme activity and substrate concentrations were observed. Chay and Cho (1982) extended the concept and computationally obtained similar results of oscillating elements of the enzyme reactions.

F.1 Horatian Oscillations in Glycolysis

Considering the earlier model of Sel'kov (see G & G), and assuming that the system is continuously stirred, Tomita and Daido (1980) obtained horatian behavior under forced conditions.

G Peroxidase Catalyzed Reactions

In analyzing oscillations in peroxidase catalyzed aerobic oxidation of NADH, Fedkina et al. (1981) experimentally obtained oscillatory results in concentrations of peroxidase compound Co(III) and O_2 and NADH. They studied the influence of temperature change on the oscillations.

H Decomposition of Sodium Dithionite

There appears to be no recently reported activity for this reaction.

I Bimolecular Model

Gillespie and Mangel (1981) proposed a stochastic formulation of chemical kinetics and presented an explanation for the limit cycle of the bimolecular model.

Jha and Prasad (1981) considered the stability analysis of this reaction, and reiterated the known results. For a concise study, see (G & G).

J Abstract Reaction Systems

The applications of these models [see G & G] are numerous in theoretical and experimental studies of known reactions, see Sections C.5, D.3, E.11, and F.1.

K Uncatalyzed Reactions

Since their introduction in the late 1970's (see G & G), uncatalyzed reactions have become the subject of numerous studies. Several systems with various substrates have been experimented with. These reactions are summarized in Table III.K., and contributions are briefly discussed in this section.

Table III.K. Uncatalyzed Reactions

Substrate	References
Pyrogallic acid	Ameta & Khandia (1982)
Various organic compounds	Chopin-Dumas (1981)
1,4-cyclohexanedione(tetrahydroquinone)	Farage & Janjic (1981-1, 2, 3)
3-alizarinsulfonic acid sodium salt	Gupta et al. (1981)
Salicylic acid ⎱ 5-sulfosalicylic acid ⎰	Gupta & Srinivasulu (1981)
3,4-dimethoxy benzaldehyde	⎧ Mittal et al. (1981) ⎨ Nair et al. (1982)
Hydroquinone	Nair et al. (1980-1, 2)
Disodium salt of 4,5-dihydroxy-1,3-bensene disulfonic acid (Tiron)	Nair et al. (1981-1)
Cathecol	Nait et al. (1981-2)
Hydroquinone and nitrophenols	Nair et al. (1981-3)
$[Ce_6[14]cis\text{-}diene]$ $(ClO_4)_2$	⎧ Yatsimirskii et al. (1980) ⎨ Kol'chinskii and Yatsimirskii (1980)

Ameta and Khandia (1982) studied kinetic oscillations in the pyrogallic acid-$KBrO_3$-H_2SO_4 system at 20 °C as a function of reactant concentrations. It was observed that the induction period for oscillations increases with an increase in pyrogallic acid concentration and decreases with increasing potassium bromate and sulfuric acid concentrations.

Chopin-Dumas (1981) proposed two criteria in selecting organic substrates that can give uncatalyzed oscillating reactions. These two criteria are based on the oxidation potentials at the anodes, and various oxidation mechanisms. Based on these two criteria, a number of organic compounds were selected and tabulated.

Farage and Janjic (1982-1 and 2) observed oscillations in the concentrations of bromide and redox potentials during the uncatalyzed oxidation of 1,4-cyclohexanedione by bromate in sulfuric, nitric (1982-1), perchloric and orthophosphoric (1982-2) acid solutions. The system does not require a catalyst such as the redox couple Ce(IV)/Ce(III) or Mn(III)/Mn(II) of the B–Z reaction. Experimenting with this system Farage and Janjic (1982-3) observed that temperature, stirring and oxygen affect the frequency or amplitude of oscillations.

Gupta et al. (1981) studied uncatalyzed and catalyzed oscillatory behavior in the redox potential during the oxidation of 3-alizarinsulfonic acid sodium salt by $KBrO_3-H_2SO_4$ and suggested a mechanism. Gupta and Srinivasulu (1981) reported a comparative study of oscillatory behavior in the redox potential in a slowly stirred closed system of salicyclic and 5-sulfosalicyclic acid (5-SSA) with acidic (H_2SO_4) bromate in the presence and absence of catalysts. Ferroin (1,10-phenanthrolineiron(II) sulfate) as catalyst enhances the oscillatory behavior of 5-SSA considerably.

Mittal, et al. (1981) reported a new reaction which consists of uncatalyzed acidic bromate oxidation of 3,4-dimethoxy benzaldehyde resulting in oscillations in redox potential. Recently, Nair, et al. (1982) reported the oscillating behavior in the uncatalyzed aqueous acidic bromate oxidation of 3,4-dimethoxy benzaldehyde.

Nair, et al. (1980-1, 2) recorded periodic oscillations of potential for a stirred closed system of hydroquinone and bromate in dilute H_2SO_4 in the presence (a B-Z reaction) or absence (uncatalyzed reaction) of a catalyst (Mn(II), Ce(IV) or ferroin). Optimum concentrations for oscillations to occur were reported and a probable mechanism was suggested. Nair, et al. (1981-1) also studied the uncatalyzed oscillatory reaction of disodium salt of 4,5-dihydroxy-1,3-benzenedisulfonic acid (Tiron)/BrO_3^-/H_2SO_4 system. The reaction was repeated under the B-Z conditions and the revival of oscillations with the introduction of a catalyst after termination of the uncatalyzed oscillations were also observed. Nair, et al. (1981-2) reported oscillations of the redox potential in the oxidation of cathechol with $KBrO_3$ in H_2SO_4 medium. However, these oscillations occurred only over a narrow range of concentrations and after a longer induction period. Further, Nair, et al. (1981-3) observed oscillations in uncatalyzed bromate oxidation of hydroquinone and nitrophenols. The catalysts of the B-Z reaction increased the number of oscillations and shortened the induction period. On the other hand, the reactions were strongly inhibited by stirring.

Yatsimirskii et al. (1980) and Kol'chinskii and Yatsimirskii (1982) found that uncatalyzed oxidation of a complex of copper with a macrocyclic ligand, Me_6[14]cis-diene, i.e. $Cu(cd)(ClO_4)_2$, oscillates. Oscillations of the redox potential using a platinum electrode and those of the Br^- concentrations using a bromide-selective electrode were recorded. While oscillations for the system $KBrO_3^-$/$Cu(cd)(ClO_4)_2$ were observed, the systems with $Cu(td)(ClO_4)_2$ and $Cu(teta)(ClO_4)_2$ did not result in oscillations, where (td) and (teta) represent the macrocyclic ligand Me_6[14]trans-diene) and Me_6[14]ane, respectively.

L Oxidation by Chlorite

During the recent years a number of studies has been made to design homogeneous oscillators systematically. While the others are the *iodate* (Bray-Liebhavsky and Briggs-Rauscher) and *bromate* oscillators (Belousov-Zhabotinskii), these new reactions are *chlorite* oscillators. For the chlorite oscillators, Orban et al. (1982-3) have also given a preliminary classification.

In this Section, these new reactions are summarized in groups as indicated by the titles of the subsections, see also Table III.L.

Table III.L. Oxidation by Chlorite

System	terms	term	References
H_3AsO_3,	IO_3^-,	ClO_2^-	De Kepper et al. (1981-1, 2)
			Gribshaw et al. (1981)
I_2,		ClO_2^-	Grant et al. (1982)
I^-,		ClO_2^-	Dateo et al. (1982)
$S_2O_3^{2-}$,	IO_3^-,	ClO_2	De Kepper et al. (1982)
(Batch oscillations)			
$Fe(CN)_6^{4-}$,	IO_3^-,	ClO_2^-	Orban et al. (1981)
SO_3^{2-},	IO_3^-,	ClO_2^-	Orban et al. (1981)
Ascorbic acid,	IO_3^-,	ClO_2^-	Orban et al. (1981)
$CH_2O \cdot HSO_2Na$,	IO_3^-,	ClO_2^-	Orban et al. (1981)
$CH_2(COOH)_2$,	IO_3^-,	ClO_2^-	Orban et al. (1981)
$Cr_2O_7^{2-}$,	I^-,	ClO_2^-	Orban et al. (1982-2)
$Fe(CN)_6^{4-}$,	I_2,	ClO_2^-	Orban et al. (1982-2)
SO_3^{2-},	I_2,	ClO_2^-	Orban et al. (1982-2)
$S_2O_3^{2-}$,	I_2,	ClO_2^-	Orban et al. (1982-2)
MnO_4^-,	IO_3^-,	ClO_2^-	Orban et al. (1982-2)
$S_2O_3^{2-}$,	iodine-free	ClO_2^-	Orban et al. (1982-1)
Horatian:			Orban and Epstein (1982-1)
Fe(II)		Uncatalyzed	Orban and Epstein (1982-2)
(No oscillations)			

L.1 Arsenite Oxidation

De Kepper, et al. (1981-1) designed a homogeneous oscillating reaction by coupling the autocatalytic oxidation of arsenite by IO_3^- to the autocatalytic ClO_2^--IO_3^- reaction in a CSTR. Both I_2 and I^- concentrations oscillate with the concentration of the latter changing by a factor of $> 10^5$ during each cycle. This arsenite-iodate-chlorite system was obtained in two separate reactions. The oxidation of arsenite by iodate, a reaction autocatalytic in *iodide* is:

(R1) $\qquad 3\,H_3AsO_3 + IO_3^- = 3\,H_3AsO_4 + I^-$

This is a *bistable* system. To couple it with a *feedback*, the second reaction, autocatalytic in *iodine* was taken:

(R2) $\qquad 4\,H^+ + ClO_2^- + 4\,I^- = 2\,H_2O + Cl^- + 2\,I_2$

Therefore to induce the oscillations *chlorite* is introduced. For a certain range of initial KIO_3 and $NaClO_2$ and As_2O_3 concentrations oscillations in I^- concentration were observed.

De Kepper et al. (1981-2) further discussed the arsenite-iodate-chlorite system. On the basis of three irreversible processes (A, B, and C) involved in the acidic oxidation of arsenite by iodate (due to Eggert, J and Scharnow, B. Z., Elektrochem. Z. (1921) 27, 455–470),

$$IO_3^- + 3\,H_3AsO_3 = I^- + 3\,H_3AsO_4 \qquad\qquad (A)$$

$$IO_3^- + 5\,I^- + 6\,H^+ = 3\,I_2 + 3\,H_2O \qquad\qquad (B)$$

$$I_2 + H_3AsO_3 + H_2O = 2\,I^- + H_3AsO_4 + 2\,H^+ \qquad\qquad (C)$$

the net stoichiometry is 2(B) + 5(C) or A + B + 2(C):

$$2\,IO_3^- + 5\,H_3NO_3 + 2\,H^+ = I_2 + 5\,H_3AsO_4 + H_2O$$

The model equations are given as:

$$dx_1/dt = -r_A - r_B + k_0(x_{10} - x_1)$$
$$dx_2/dt = r_A - 5r_B + 2r_C + k_0(\alpha x_{10} - x_2)$$
$$dx_3/dt = 3r_A - r_C - (k_s + k_0)\,x_3$$
$$dx_4/dt = -3r_A - r_C + k_0(x_{40} - x_4)$$

where $x_1 = IO_3^-$, $x_2 = I^-$, $x_3 = I_2$, $x_4 = H_3ASO_3$ and the rates are

$$r_A = k_A x_1 x_4$$
$$r_B = k_{B1}[H^+]^2\, x_1 x_2 + K_{B2}[H^+]^2\, x_1 x_2^2$$
$$r_C = k_C x_4 x_3/[H^+]\, x_2$$

Here k_0 is the reciprocal of the residence time τ and α is the fraction of iodide impurity in the iodate. The rate constant k_s is the associated first-order (iodine decay rate) rate constant. The six free parameters are k_A, k_{B1}, k_{B2}, k_C, k_s, α and the constraints are k_0, x_{10}, x_{40}, and $[H^+]$.

Gribshaw et al. (1981) studied the chemical waves in the acidic iodate oxidation of arsenite.

L.2 Iodine Oxidation

Grant et al. (1982) studied the oxidation of iodine by chlorite ion. The stoichiometry of the reaction between ClO_2^- and I_2 at pH 2–5 and low I^- was determined to be $5\,ClO_2^- + 2\,I_2 + 2\,H_2O \rightarrow 5\,Cl^- + 4\,IO_3^- + 4\,H^+$. The kinetics of this reaction were studied by flow spectrometry at variable ionic strength. A mechanism was proposed involving formation of the key intermediate, $IClO_2$ by reaction between ClO_2^- and I_2, I_2OH^- and IOH_2^+. The rate constants were identified with the reaction $I_2 + H_2O \rightarrow IOH_2^+ + I^-$.

L.3 New Reducing Agents

Orban et al. (1981) studied reactions where *chlorite* and *iodate* are common. However a wide variety of *reducing agents* or substrates in place of *arsenite* could be used. One electron (ferrocyanide) and two-electron (thiosulphate) reductants react with $NaClO_2$ and KIO_3 in a CSTR. The substrates giving oscillations in these systems, in addition to H_3AsO_3 are ascorbic acid, $K_4[Fe(CN)_6]$, Na_2SO_3, $Na_2S_2O_3$, $CH_2O \cdot HSO_2Na$ and malonic acid with KI. Oscillations in iodide and redox potential were observed.

L.4 Heterogeneous (Bromate-Hydrogen-Platinum-Acid) System

In a B-Z type reaction, where the organic substrate is replaced by hydrogen and catalyst is replaced by platinum, Orban and Epstein (1981) observed oscillations in potential. The oscillatory process, however, takes place only on the surface of the platinum catalyst, analogous to the carbon monoxide (Keil, and Wicke (1980)) and ethylene oxidation (Vayenas et al. (1980)) at platinum surfaces. The reagents are H_2SO_4, $KBrO_3$, [or KIO_3, or $NCIO_2$]. Oscillations in potential were recorded following

1) switching off of the stirrer,
2) adding Ce(IV) as an initiator,
3) adding Vanadium(IV) as an initiator.

In the iodate-hydrogen-platinum-acid system the oscillatory range was found to be narrow and sensitive to physical parameters. Under similar conditions, with anions such as $Cr_2O_7^{2-}$ and MnO_4^- no oscillations were recorded.

L.5 Autocatalytic Chlorite-Iodate-Reaction in a CSTR

Dateo et al. (1982) studied the reaction between chlorite and iodide in acidic solution in a CSTR over a range of flow rates, pH, and input chlorite and iodide concentrations. The system was found to exhibit bistability, and at high $[ClO_2^-]$ and $[I^-]$, sustained oscillations. The key reaction which gives rise to these phenomena is $ClO_2^- + 4 I^- + 4 H^+ = Cl^- + 2 I_2 + 2 H_2O$, which is autocatalytic in iodine and is inhibited by iodide. The existence of bistability and oscillations in this system with the "cross-shaped phase diagram" model of Boissonade and De Kepper (1980) was illustrated. Possible topological configurations for various flow rates and k_0 were discussed.

L.6 Chlorite-Thiosulphate Reaction without Iodine Species

Orban et al. (1982-1) showed that the reaction between chlorite and thiosulphate in a CSTR exhibits oscillations. The oscillations in the potential of platinum redox electrode were recorded. These periodic oscillations showed different patterns as the input concentrations of chlorite and thiosulfate as well as pH (between 2 and 5) and flow rate varied.

Orban and Epstein (1982-1) studied the same system and reported horatian behavior in potential for various regions on the plane of chlorite and thiosulphate concentrations.

L.7 Batch Oscillations

De Kepper et al. (1982) found batch oscillations (oscillations in the absence of flow) of redox or I^- sensitive electrode in reactions:

A) ClO_2^-, IO_3^-, I^-, $CH_2(COOH)_2$
B) ClO_2^-, I^-, $CH_2(COOH)_2$, H_2SO_4
C) ClO_2^-, IO_3^-, $S_2O_3^{2-}$, H_2SO_4

Also they observed spatial wave patterns in a reaction with

$CH_2(COOH)_2$, NaI, $NaClO_2$, H_2SO_4 and starch.

L.8 Additional Studies

Orban and Epstein (1982-1) studied the autocatalytic oxidation of Fe(II) in nitric acid in a CSTR. In this preliminary work, although the possibility still remains, they did not observe oscillations.

M Miscellaneous Studies

Hatami (1981) determined the dynamic behavior of the total combustion system theoretically for conditions with and without flame and presented an experimental procedure. Theoretical calculations were verified experimentally. It is found that the recirculation parameter, the gas chamber volume and the time lag influence the instability of the system resulting in variations in the oscillatory behavior of the pressure.

Toby and Ulrich (1980) studied oxidation of CO by ozone and under certain experimental conditions observed undamped oscillations in chemiluminescence.

Young, et al. (1982) experimented with oxidation of ascorbic acid (AA) by molecular oxygen, and Cu(II) as catalyst. Concentration of AA was found to oscillate. A mathematical model was also given.

Ganapathisubramanian and Noyes (1981) studied the decomposition of hydrogen peroxide using the following systems:

1) H_2O_2, H_2SO_4 and Fe(III) (as sulfate),

2) H_2O_2, H_2SO_4 and Fe(II) as sulfate,

and obtained oscillations in the pressure of O_2.

Laplante and Pottler (1982) observed fluorescence oscillations upon irradiating the 9,10-dimethyl anthracene/chloroform systems continuously. Fluorescence oscillations were found to occur immediately after the reaction was started. However, initial oscillations were usually found to be of horatian nature and periodic oscillations were obtained following a horatian initial period. In some cases the periodic behavior was not reproducible even though the nonperiodic oscillatory behavior could be reproduced. Experiments were carried out to determine the effect of e.g., excitation wavelength, substrate concentration and specificity. The mechanism has not yet been elucidated.

Oscillations in chemical systems in general are monitored by ion selective electrodes. In some cases the interpretations of potential measurements are contradictory. Noszticzius et al. (1981) studied the potential responses of halide ion selective electrodes under the experimental conditions of some known oscillating systems. They concluded that in the presence of hypohalous acids the potential response below the solubility limit of halide ions is due to only the hypohalous acid present and not to halide concentration.

Experimenting with hydrolysis of peptides, Slobodyanikova et al. (1980) concluded that oscillations may be observed. The character of these oscillations were found to be affected by the substrate/catalyst ratio. Oscillations in amine nitrogen during hydrolysis of peptide and amino acid mixture were recorded.

The reaction of CaS in an oxidizing atmosphere was studied by Lynch and Elliott (1980). Calcium sulfide pellets were reacted in $Ar-O_2$ mixture with oxygen. Oxidation of CaS to CaO and the decomposition of $CaSO_4$ lead to oscillatory behavior at certain temperatures and partial pressures of oxygen.

Tomashov et al. (1981) working with chromium alloys with Ru, Os and Ir recorded oscillations in the electric potentials.

N General Models and Mathematical Techniques

N.1 Two Dimensional Models

Alekseev and Kol'tsov (1982) analyzed a three step catalytic reaction where two intermediates (x, y) were considered. The model was given as two differential equations with product nonlinearity in x and y.

Ivanov et al. (1980) modelled a class of Langmuir-Hinshelwood reactions, and by analyzing the mathematical model given in two dimensions, they obtained limit cycle oscillations and showed that the influence of adsorbed species on the catalytic reaction rate may lead to periodic oscillations.

For an open enzymatic reaction with an enzyme subject to continuous synthesis and breakdown, Nazarenko and Sel'kov (1981) gave a two-dimensional model. Furthermore, Sel'kov and Nazarenko (1981) proposed a model with an external perturbation and plotted oscillatory solutions.

N.2 Three Dimensional Models

Starting with a three dimensional model of carbohydrate metabolism, Taranenko (1981) analyzed the periodic behavior of the system by varying the parameters, and showed the existence of period doubling and horatian oscillations.

Iwamoto and Seno (1980) proposed a three dimensional mathematical model and analyzed it under the influence of external fluctuation terms. In (1982) Iwamoto et al. gave a three dimensional model to explain small and large amplitude oscillations observed in the reactions.

Decroly and Goldbeter (1981) modelled an enzyme reaction with two positive feedbacks, as a three dimensional model of differential equations, and for various values of the parameter K_s, the first order rate constant, they obtained single or two limit cycles as well horatian behavior.

N.3 Additional Studies

Heineman and Poore (1981) developed a numerical technique to analyze tubular reactor systems.

Another study on numerical analysis methods for stiff differential equations of chemical reactions, including the B-Z reaction as an example, was reported by Prokopakis and Seider (1981).

Lynch and Wanke (1981) showed that simplifying assumptions of models result in loss of the true behavior of the system modelled.

Selegny and Vincent (1980) considered a four enzyme system and proposed a formal analysis of such systems.

Theoretical studies for the B-Z system and the solid catalyzed reactions are included in Sections C.2 and E.10, respectively.

Demet Gurel and Okan Gurel

IV References

References are listed in alphabetical order. The notations on the left hand column designate (Section, Part). Review articles are denoted by letter R. Brooks are indicated by letter B.

(IIIC)   Adamcikova, L., Sevcik, P.: Heterogeneous Oscillation Reaction in Tartaric Acid —
1982-1   KBrO$_3$—MnSO$_4$—H$_2$SO$_4$ System, Collect. Czech. Chem. Commun. 47, 2333–2335 (Eng.)

(IIIC)   Adamcikova, L., Sevcik, P.: A Completely Inorganic Oscillating System of the Belousov-
1982-2   Zhabotinskii Type, Int. J. Chem. Kinet. 14 (7) 735–738

(IIIC)   Adamcikova, L., Treindl, L.: Polarographic Study of the Kinetics and Mechanism of
1976   Belousov-Zhabotinskii Reaction, Collect. Czech. Chem. Commun. 41, 3521–3527 (Eng.)

(IIIE)   Adlhoch, W., Lintz, H. G., Weisker, T.: Oscillations of Reaction Rate During the
1981   Reaction of Nitric Oxide with Carbon Monoxide under Knudsen Conditions, Surf. Sci.
            103, 576–585. (Ger.)

(IIIN)   Alekseev, B. V., Kol'tsov, N. I.: Self-Oscillation Conditions of a Nonlinear Three-Step
1982   Catalytic Reaction. React. Kinet. Catal. Lett. 19 (1-2) 15–21 (Russ.)

(IIIK)   Ameta, S. C., Khandia, B. L.: Studies in Oscillatory Behavior of Oxidation of Pyrogallic
1982   Acid by Potassium Bromate, Z. Phys. Chem. (Leipzig) 263, 410–412

(IIIC, D)   Bar-Eli, K., Geiseler, W.: Mixing and Relative Stabilities of Pumped Stationary States,
1981   J. Phys. Chem. 85 (23) 3461–3468

(IIIB)   Betteridge, D., Joslin, M. T., Lilley, T.: Acoustic Emissions from Chemical Reactions.
1981   Anal. Chem. (Wash.) 53(7) 1064–1073

(R)   Boiteux, A., Hess, B., Sel'kov, E. E.: Creative Functions of Instability and Oscillations
1980   in Metabolic Systems, Curr. Top. Cell. Regul. (B. L. Horecker & E. R. Stadtman, eds.)
            17, 171–203

(IIIC)   Bolletta, F., Balzani, V.: Oscillating Chemiluminescence from the Reduction of Bromate
1982   by Malonic Acid Catalyzed by Tris(2,2-bipyridine) ruthenium (II), J. Am. Chem. Soc.
            104(15) 4250–4251

(IIIE)   Bond, J. R., Gray, P., Griffiths, J. F., Scott, S. K.: Oscillations, Glow and Ignition in
1982   Carbon Monoxide Oxidation. II. Oscillations in the Gas-Phase Reaction in a Closed
            System, Proc. R. Soc. London (Ser.) A 381, 293–314

(IIIC)   Botre, C., Lucarini, C., Memoli, A.: On the Entropy Production in Oscillating Chemical
1981   Systems, Bioelectrochem. Bioenerg. 8, 201–212. (A section of J. Electroanal. Chem. v. 128)

(IIIE)   Bouillon, M., Dagonnier, R., Dufour, P., Dumont, M.: Comments on "Isothermal
1982   Sustained Oscillations in a very simple Surface Reaction" by C. G. Takoudis, L. D.
            Schmidt and R. Aris, Surf. 115(2), L111–L116

(R)   Brisset, J. L.: Oscillating Chemical Systems, Bull. Union Physiciens 75(629) 371–382
1980   (Fr.)

(IIIC)   Burger, M., Kőrős, E.: Chemistry of the Belousov-Zhabotinskii Oscillatory Systems.
1980-1   III. Role of Bromomalonic Acid. Magy. Kem. Foly 86, 8–14

(IIIC)   Burger, M., Kőrős, E.: Prerequisite for Oscillatory Behavior in the Malonic Acid,
1980-2   Bromate and Catalyst Reacting Systems, Ber. Bunsenges. Phys. Chem. 84(4) 363–366.

(IIID)   Caprio, V., Insola, A., Lignola, P. G.: Isobutane Cool Flames in CSTR: The Behavior
1981   Dependence on Temperature and Residence Time, Combust. Flame 43(1) 23–33

(IIIE)   Chang, H. C., Aluko, M.: Comments on the Mode*j* for Isothermal Oscillations of
1982   Ethylene Oxidation on Platinum, J. Catal. 73(1) 198–200

(IIIF)   Chay, T. R.: A Model for Biological Oscillations. Proc. Natl. Acad. Sci. U.S.A. 78,
1981   2204–2207

(IIIF)   Chay, T. R., Cho, S. H.: On Exploring the Basis for Slow and Fast Oscillations in
1982   Cellular Systems, Biophys. Chem. 15(1) 9–13

108

(IIIK)   Chopin-Dumas, J.: Periodic Reactions with Bromate, I. Selection Criteria for Organic
1981   Substrates in Uncatalyzed Oscillatory Reaction, J. Chim. Phys. Phys.-Chim.-Biol. 78(5)
              461–469. (Fr.)

(IIIE)   Chumakov, G. A., Slin'ko, M. G., Belyaev, V. D.: Complex Changes in Heterogeneous
1980   Catalytic Reaction Rates. Dokl. Akad. Nauk SSR. 253, 653–658. (Phys. Chem.) (Rus)
              [Dokl. Physical Chem. 253, 637–641 (Eng. trans.)]

(IIIA)   Cooke, D. O.: On the Effect of Copper (II) and Chloride Ions on the Iodate-Hydrogen
1980-1   Peroxide Reaction in the Presence and Absence of Manganese (II), Int. J. Chem. Kinet. 12,
              671–681

(IIIA)   Cooke, D. O.: The Hydrogen Peroxide-Iodic Acid-Manganese (II)-Acetone Oscillating
1980-2   System: Further Observations. Int. J. Chem. Kinet. 12, 683–98

(IIID)   Crooke, P. S., Wei, C. J., Tanner, R. D.: Effect of the Specific Growth Rate and Yield
1980   Expressions on the Existence of Oscillatory Behavior of a Continuous Fermentation
              Model. Chem. Eng. Commun. 6(6) 333–347

(IIIC)   D'Alba, F., Serravalle, G.: Oxidation of Malonic Acid by Potassium Bromate (Reaction
1981   of Zhabotinskii). Comparison between Different Catalysts by Electrochemical Method,
              J. Chim. Phys. Phys. Chim. Biol. 78, 131–134

(IIIL)   Dateo, C. E., Orban, M., De Kepper, P., Epstein, I. R.: Systematic Design of Chemical
1982   Oscillators, Part 5. Bistability and Oscillations in the Autocatalytic Chlorite-Iodide
              Reaction in a Stirred-Flow Reactor, J. Am. Chem. Soc. 104(2) 504–509

(IIIN)   Decroly, O., Goldbeter, A.: Birhytmicity, Chaos and other Patterns of Temporal Self-
1982   Organization in a Multiply Regulated Biochemical System, Proc. Natl. Acad. Sci. U.S.A.
              79, 6917–6921

(IIIC, D)   De Kepper, P., Boissonade, J.: Theoretical and Experimental Analysis of Phase Diagrams
1981   and Related Dynamical Properties in the Belousov-Zhabotinskii System, J. Chem.
              Phys. 75, 189–195

(IIIB)   De Kepper, P., Epstein, I. R.: Mechanistic Study of Oscillations and Bistability in the
1982   Briggs-Rauscher Reaction, J. Am. Chem. Soc. 104(1) 49–55

(IIIL)   De Kepper, P., Epstein, I. R., Kustin, K.: Systematic Design of Chemical Oscillators
1981-1   Part 2. A Systematically Designed Homogeneous Oscillatory Reaction: The Arsenite-
              Iodate-Chlorite System, J. Am. Chem. Soc. 103, 2133–2134

(IIIL)   De Kepper, P., Epstein, I. R., Kustin, K.: Systematic Design of Chemical Oscillators
1981-2   Part 3. Bistability in the Oxidation of Arsenite by Iodate in Stirred Flow Reactor,
              J. Am. Chem. Soc. 103, 6121–6127

(IIIL)   De Kepper, P., Epstein, I. R., Kustin, K., Orban, M.: Systematic Design of Chemical
1982   Oscillators. Part 8. Batch Oscillations and Spatial Wave Patterns in Chlorite Oscillating
              Systems, J. Phys. Chem. 86(2) 170–171

(IIIB)   Dutt, A. K., Banerjee, R. S.: Iodine Clock Oscillating Reaction, J. Indian Chem.
1980   Soc. 57, 751–753.

(IIIB)   Dutt, A. K., Banerjee, R. S.: Briggs-Rauscher Oscillating Reaction Using New Com-
1981-1   pounds. J. Indian Chem. Soc. 58, 546–549

(IIIB)   Dutt, A. K., Banerjee, R. S.: Oscillating Iodine Clock Reaction in Presence of Chloride
1981-2   Ions. J. Indian Chem. Soc. 58, 717–719

(IIIB)   Dutt, A. K., Banerjee, R. S.: Studies on Kinetic Parameters of Briggs-Rauscher
1982   Oscillating Reaction, Z. Phys. Chem. (Leipzig) 263, 298–304

(B)   Ebert, K. H., Deuflhand, P. and Jager, W. (Eds.): Modelling of Chemical Reaction
1981   Systems, Proc. an Intern. Workshop, Heidelberg, Fed. Rep. Germany, September 1–5,
              1980. Springer Series in Chemical Physics. v. 18

(IIIC)   Edelson, D.: Mechanistic Details of the Belousov Zhabotinskii Oscillations IV. Sen-
1981   sitivity Analysis, Int. J. Chem. Kinet. 13, 1175–1189

(IIIE)   Ertl, G., Norton, P. R., Ruestig, J.: Kinetic Oscillations in the Platinum-Catalyzed
1982   Oxidation of Cobalt. Phys. Rev. Lett. 49, 177–180

109

(IIIC)   Farage, V. J., Janjic, D.: Calorimetric Study of the Bromate/Cerium IV/Cyclohexanone
1980-1   and Bromate/Cerium (IV)/Cyclopentanone Oscillating Systems, React. Kinet. Catal.
           Lett. 15, 487–491 (Eng.)

(IIIC)   Farage, V. J., Janjic, D.: 43. Oscillating Chemical Reactions, III. The Effects of the
1980-2   Temperature and of the Chemical Composition on the "Induction Period" of the Systems
           BrO_3^-/Ce(IV)/Cyclohexanone and BrO_3^-/Ce(IV)/Cyclopentanone, Helvetica Chimia Acta
           63, 433–437. (Eng.)

(IIIC)   Farage, V. J., Janjic, D.: Effect of Mechanical Agitation on Oscillating Chemical
1981   Reactions in the Presence of Oxygen or Nitrogen, Chimia 35, 289–291 (Fr.)

(IIIK)   Farage, V. J., Janjic, D.: Uncatalyzed Oscillatory Chemical Reaction. Oxidation of
1982-1   1,4-Cyclohexanedione by Bromate in Sulfuric or Nitric Solution, Chem. Phys. Lett.
           88(3) 301–304.

(IIIK)   Farage, V., Janjic, D.: Uncatalyzed Oscillatory Chemical Reaction. Oxidation of
1982-2   1,4-Cyclohexanedione by Bromate in Perchloric or Orthophosphoric Acid Solution.
           Inorg. Chem. Acta Lett. 65(2) L33–L34.

(IIIK)   Farage, V. J., Janjic, D.: Uncatalyzed Oscillatory Chemical Reactions, Effect of Different
1982-3   "Constraints" during the Oxidation Reaction of 1,4-Cyclohexanedione by Acidic Bro-
           mate. Chem. Phys. Lett. 93(6) 621–624.

(IIIG)   Fed'kina, V. R., Bronnikova, T. V., Ataullakhanov, F. I.: Slow Oscillations in Peroxidase-
1981   Oxidase Reactions, Stud. Biophys. 82(3) 159–164 (Eng.)

(IIIC)   Fujisaka, H., Yamada, T.: Limit Cycles and Chaos in Realistic Models of the Belousov-
1980   Zhabotinskii Reaction System. Z. Phys. B. 37, 265–75.

(IIIB)   Furrow, S. D.: Briggs-Rauscher Oscillator with Methylmalonic Acid. J. Phys. Chem. 85,
1981   2026–2031.

(IIIB)   Furrow, S. D.: Iodine-Production Subsystem of the Briggs-Rauscher Oscillating Reac-
1982   tion. Effect of Crotonic Acid. J. Phys. Chem. 86, 3089–3094.

(IIIB)   Furrow, S. D., Noyes, R. M.: The Oscillatory Briggs-Rauscher Reaction — 1. Exami-
1982-1   nation of Subsystems. J. Am. Chem. Soc. 104(1) 38–42.

(IIIB)   Furrow, S. D., Noyes, Richard, M.: The Oscillatory Briggs-Rauscher Reaction —
1982-2   2. Effects of Substitutions and Additions. J. Am. Chem. Soc. 104(1) 42–45.

(IIIM)   Ganapathisubramanian, N., Noyes, R. M.: Chemical Oscillstions and Instabilities.
1981   43. Oscillatory Oxygen Evolution during Catalyzed Disproportionation of Hydrogen
           Peroxide. J. Phys. Chem. 85, 1103–1105.

(IIIC)   Ganapathisubramanian, N., Noyes, R. M.: Chemical Oscillations and Instabilities.
1982-1   48. Reliability of Bromide Ion-Selective Electrodes for Studying the Oscillatory Belouzov-
           Zhabotinskii Reaction. J. Phys. Chem. 86(16) 3217—3222.

(IIIC)   Ganapathisubramanian, N., Noyes, R. M.: Chemical Oscillations and Instabilities.
1982-2   49. Bromate-Driven Oscillators in the Presence of Excess Silver Ion. J. Phys. Chem. 86,
           5155–5157.

(IIIC)   Ganapathisubramanian, N., Noyes, R. M.: Chemical Oscillations and Instabilities.
1982-3   50. Additional Complexities during Oxidation of Malonic Acid in the Belousov-Zhabo-
           tinskii Reaction. J. Phys. Chem. 86, 5158–5162.

(IIIC)   Ganapathisubramanian, N., Noyes, R. M.: A Discrepency between Experimental and
1982-4   Computational Evidence for Chemical Chaos. J. Chem. Phys. 76, 1770–1774.

(IIIC, D) Geiseler, W.: Sustained Oscillations of an Autocatalytic Reaction in a Stirred Flow
1982-1   Reactor. Ber. Bunsenges. Phys. Chem. 86, 721–724.

(IIIC, D) Geiseler, W.: Multiplicity, Stability and Oscillations in the Stirred Flow Oxidation
1982-2   of Manganese(II) by Acidic Bromate. J. Phys. Chem. 86(22) 4394–4399.

(IIIC, D) Geiseler, W., Bar-Eli, K.: Multistability in Flow Systems of the Belousov-Zhabotinskii
1981   and Related Systems. Springer Ser. Chem. Phys. 18, 268–274.

(III I)   Gillespie, D. T., Mangel, M.: Conditioned Averages in Chemical Kinetics. J. Chem.
1981   Phys. 75(2) 704–709.

(IIIF)   Goldbetter, A., Venieratos, D.: Analysis of the Role of Enzyme Cooperativity in
1980   Metabolic Oscillations. J. Mol. Biol. 138, 137–144

(IIIL)   Grant, J. L., De Kepper, P., Epstein, I. R., Kustin, K., and Orban, M.: Systematic
1982   Design of Chemical Oscillators. Part 9. Kinetics and Mechanism of the Oxidation of
            Iodine by Chlorite Ion, Inorganic Chem. 21(6) 2192–2196

(IIIL)   Gribshaw, T. A., Showalter, K., Banville, D. L., Epstein, I. R.: Chemical Waves in Acidic
1981   Iodate Oxidation of Arsenite. J. Chem. Phys. 85, 21252–2155

(IIIK)   Gupta, V. K., Nair, P. K. R., Srinivasulu, K.:, Oscillatory Behavior of the Reaction
1981   of 3-Alizarinsulfonic Acid Sodium Salt with Acidic Bromate. React. Kinet. Catal. Lett.
            18(1-2) 45–49.

(IIIK, C)   Gupta, V. K., Srinivasulu, K.: Uncatalyzed and Catalytic Bromate Driven Oscillations
1981   in Salicylic and 5-Sulfosalicylic Acid. React. Kinet. Catal. Lett. 19(1-2) 193–196.

(IIIC)   Hebashi-Krayenbuhl, D., Janjic, D.: Belousov-Zhabotinskii type Chemical Oscillator
1982   Involving Phosphonoacetic Acid. Chimia 36(3) 123–124. (Fr.)

(III)   Habon, I., Kőrős, E.: Investigation on the Pyrogallol-Bromate-Acid-Oscillatory Chemical
1979   System. Ann. Univ. Sci. Budap. Rolanlo Eotvos Naminatae, Sect. Chim. 15, 23–29
            (Eng.)

(IIIC)   Handlirova, M., Tockstein, A.: New Catalysts for the Belousov-Zhabotinskii Reaction.
1980   Collect. Czech. Chem. Commun. 45(10) 2621–2623 (Eng.)

(IIIM)   Hatami, R.: Calculation of Combustion Instabilities of Enclosed Diffusion Flames.
1981   Chem. Eng. H, (Lausanne) 22(1) 1–14

(IIIE)   Hegedus, L. L., Chang, C. C., McEwen, David, J., Sloan, E. M.: Response of Catalyst
1980   Surface Concentrations to Forced Concentrations Oscillations in the Gas Phase. The
            NO, CO, O_2 System over Pt-Alumina. Ind. Eng. Chem. Fundam. 19(4) 367–373

(IIIN)   Heinemann, R. F., Poore, A. B.: Multiplicity Stability, and Oscillatory Dynamics of the
1981   Tubular Reactor. Chem. Eng. Sci. 36(8) 1411–1419

(IIIE)   Hlavacek, V., Rathousky, J.: Oscillatory Behavior of Metallic Honey Comb Catalysts,
1982   Chem. Eng. Sci. 37, 375–380

(IIIB)   Horsthemke, W.: Nonequilibrium Transitions Induced by External White and Colored
1980   Noise, In: Dynamics of Synergetic Systems, H. Haken (Ed.). Springer Ser. Synergetics 6,
            67–77

(IIID)   Hugo, P., Wirges, H. P.: Approximation to Calculate Self-Sustained Oscillations of a
1980   Stirred Tank Reactor. Ind. Eng. Chem. Fund. 19(4) 428–435

(IIIC, D)   Hudson, J. L., Mankin, J. C.: Chaos in the Belousov-Zhabotinskii Reaction. J. Chem.
1981   Phys. 74, 6171–6177

(IIIN)   Ivanov, E. A., Chumakov, G. A., Slinko, M. G., Bruns, D. D., Luss, D.: Isothermal
1980   Sustained Oscillations Due to the Influence of Adsorbed Species on the Catalytic Reaction
            Rate. Chem. Eng. 35, 795–803

(IIIN)   Iwamoto, K., Kawauchi, S., Sawada, K., Seno, M.: Chemical Systems Exhibiting Com-
1982   plicated Temporal Oscillations, Bull. Chem. Soc. Japan 55, 423–426

(IIIN)   Iwamoto, K., Seno, M.: Effects of Fluctuation on Dissipative Structures, III. Analysis
1980   of a Model Sensitive to Changes in External Constraint. J. Chem. Phys. 72, 4235–41

(IIIC, D)   Iwamoto, K., Seno, M.: On Behaviors of a New Chemical Reaction Model Showing
1981   Hard Oscillations. Bull. Chem. Soc. Japan 54, 669–673

(IIIE)   Jaeger, N. I., Plath, P. J., van Raaij, E.: Chemical Oscillations of the Methanol Oxidation
1981   on a Supported Palladium Catalyst. Z. Naturforschung 36A(4) 395–402. (Ger.)

(IIIE)   Jensen, K. F., Ray, W. H.: A Microscopic Model for Catalytic Surfaces, 1. Catalytic
1980-1   Wires and Gauges. Chem. Eng. Sci. 35(12) 2439–2457

(IIIE)   Jensen, K. F., Ray, W. H.: A New Virew of Ignition, Extinction and Oscillations on
1980-2   Supported Catalytic Surfaces. Chem. Eng. Sci. 35(1-2), Chem. Reac. Eng. V. 1 Contributed
            Papers, Int. Symp. on Chem. React. Eng. 6th Nice, France March 25–27, 1980, 241–248

(IIIE)   Jensen, K. F., Ray, W. H.: A Microscopic Model for Catalytic Surfaces, II. Supported
1982-1   Catalysts. U. of Wisconsin MRC Tech. Report No. 2338

(IIIE)   Jensen, K. F., Ray, W. H.: Bifurcation Behavior of Tubular Reactors. Chem. Eng.
1982-2   Sci. 37, 199–222

(IIII)   Jha, P. N., Prasad, R. K.: Stability and Phase Plane Analysis of the "Brusselator".
1981   J. Indian Chem. Soc. 58, 377–381

(IIIC)   Jorne, J.: Role of Diffusion in Trigger Wave Propagation in the Belousov-Zhabotinskii
1980   Reaction. J. Am. Chem. Soc. 102, 6196–6198

(IIID)   Kahlert, C., Rössler, O. E., Varma, A.: Chaos in a Continuous Stirred Tank Reactor
1981   with Two Consecutive First-Order Reactions, One Exo- One Endothermie. In: Ebert,
            et al. (1981) 355–365

(IIIE)   Kasemo, B., Keck, K. E., Hoegberg, T.: A Fast Response, Local Gas-Sampling System
1980   for Studies of Catalytic Reactions at $1–10^3$ Torr; Application to the Hydrogen-Deuterium
            Exchange and Hydrogen Oxidation Reactions on Platinum. J. Catal. 66(2) 441–450

(IIIE)   Keil, W., Wicke, E.: Kinetic Instabilities in Carbon Monoxide Oxidation on Platinum
1980   Catalysts. Ber. Bunsenges., Phys. Chem. 84, 377–383 (Ger.)

(IIIC)   Keszthelyi, C. P., Soos, J., Janossy, A. G. S., Kovacs, K.: Iron Catalyst Local Concen-
1981   tration Fluctuations Found in Dissipative Reactions by X-Ray Microanalysis. Bull.
            Chem. Soc. Japan 54, 321–322.

(R)   Klonowski, W.: Oscillatory Phenomena of the Dissipative Structure Type in Model
1980   Enzyme Systems. Zagadnienia Biofiz. Wspolczesnej 5, 199–230 (Polish)

(IIIK)   Kol'chinskii, A. G., Yatsimisrskii, K. B.: A New Noncatalytic Oscillating Reaction with
1980 $[Cu(Me_6[14]cis-diene)]$ $(ClO_4)_2$. Teor. Eksp. Khim. 16(4) 525–529, (Rus.). Theor. Exper.
            Chem. 16(4) 407–411 (Engl. trans.).

(IIIC)   Kőrös, E., Varga, M.: A Quantitative Study of the Iodide-Induced High-Frequency
1982   Oscillations in Bromate-Malonic Acid-Catalyst Systems. J. Phys. Chem. 86, 4839–4843.

(IIIC)   Kovalenko, A. S., Tikhonova, L. P.: Use of Inert Electrodes for Recording Concentration
1980   Oscillations in Belousov-Zhabotinskii Reactions. Teor. Eksp. Khim. 16(3) 409–415.
            (Rus.). [Theor. Exper. Chem. 16(3) 317–321 (Engl. trans.)]

(IIIC)   Kovalenko, A. S., Tikhonova, L. P., Roizman, O. M., Protopopov, E. V.: Charac-
1980   teristics of the Concentration Ranges of Existence of Oscillations in the Belousov-
            Zhabotinskii Reactions. Teor. Eksp. Khim. 16(1) 46–52, (Rus.). [Theor. Exper. Chem.
            16(1) 38–43 (Engl. trans.)]

(IIIC)   Kovalenko, A. S., Tikhonova, L. P., Yatsimirskii, K. B.: Study of Combined Action of
1981   Cerium (III, IV) Ions and Ferroin in the Oscillating Belousov-Zhabotinskii Chemical
            Reaction Teor. Eksp. Khim. 17 (4) 493–499, (Rus.). [Theor. Exper. Chem. 17 (4)
            382–387 (Engl. trans.)]

(IIIC)   Kuhnert, L., Pehl, K. W.: Oscillations in the Belousov-Zhabotinskii System (BZR).
1981-1   Catalyzed by bis-bipyridine-silver complexes. Chem. Phys. Lett. 84 (1) 155–158

(IIIC)   Kuhnert, L., Pehl, K. W.: Bipyridine Complexes of Osmium and Chromium Catalyzing the
1981-2   Oscillating Reaction between Bromate and Malonic Acid. Chem. Phys. Lett. 84 (1)
            159–162.

(IIIE)   Kurtanjek, Z., Sheintuch, M., Luss, D.: Surface State and Kinetic Oscillations in the
1980-1   Oxidation of Hydrogen on Nickel. J. Catal. 66, 11–27

(IIIE)   Kurtanjek, Z., Sheintuch, M., Luss, D.: Reaction Rate Oscillations During the Oxidation
1980-2   of Hydrogen on Nickel. Ber. Bunsenges, Phys. Chem. 84(4) 374–377

(IIIM)   Laplante, J. P., Pottler, R. H.: Study of the Oscillatory Behavior in Irradiated 9,10-
1982   Dimethyl anthracene/Chloroform Solutions. J. Phys. Chem. 86, 4759–4766

(IIIE)   Liao, P. C., Wolf, E. E.: Self-Sustained Oscillations During Carbon Monoxide Oxidation
1982   on a Platinum/γ-Aluminum Oxide Catalyst. Chem. Eng. Commun. 13(4–6), 315–326

(IIIA)   Liebhafsky, H. A., Furuichi, R., Roe, G. M.: Reactions Involving Hydrogen Peroxide,
1981   Iodine, and Iodate Ion. 7. The Smooth Catalytic Decomposition of Hydrogen Peroxide.
            J. Am. Chem. Soc. 103, 51–56

(IIID)   Lignola, P. G., Caprio, V., Insola, A., Mondini, G.: Thermokinetic Oscillations in
1980   Propane Oxidation. Ber. Bunsenges. Phys. Chem. 84(4) 369–373

(IIIM)   Lynch, D. C., Elliott, J. F.: Analysis of the Oxidation Reactions of CaS. Metall. Trans. B
1980   11B(3) 415–425

(IIIE)   Lynch, D. T., Wanke, S. E.: Examination of a Model for Oscillating Heterogeneously
1981   Catalyzed Reactions. Can J. Chem. Eng. 59(6) 766–770

(IIIF)   Malkova, I. A., Popova, S. V., Sel-Kov, E. E.: Multiplicity of Steady States and Auto-
1980   oscillations in an Open Reaction Catalyzed by Phosphofructokinase from Escherichia
            Coli. Quantitative Model. Biofizika. 25(3) 503–507. [Biophysics 25(3) 520–525 (Eng.
            trans.)]

(IIIC, D)   Maselko, J.: Experimental Studies of Complicated Oscillations. The System Manganese
1980-1   (2+) ion-malonic acid-potassium bromate-sulfuric acid. Chem. Phys. 51(3) 473–480
            (Eng.)

(IIIC, D)   Maselko, J.: Experimental Study of the Bifurcation Diagram in the Belousov-Zhabotinskii
1980-2   Reaction, React. Kinet. Catal. Lett. 15, 197–201

(IIIC)   Maselko, J.: Experimental Studies of Chaos-Type Reactions. The System Manganese
1980-3   (2+) Ion-Oxalacetic Acid-Sulfuric Acid-Potassium Bromate. Chem. Phys. Lett. 73,
            194–198

(IIIC, D)   Maselko, J.: Determination of Bifurcation in Chemical Systems. An Experimental Method.
1982   Chem. Phys. 67(1) 17–26

(IIIK)   Mittal, A., Nair, P. K. R., Srinivasulu, K.: A Novel Type of Chemical Oscillator.
1981   Int. J. Chem. Kinet. 13(3) 321–322

(IIIE)   Morton, W., Goodman, M. G.: Parametric Oscillations in Simple Catalytic Reaction
1981-1   System. Trans. Inst. Chem. Eng. 59(4) 253–259

(IIIE)   Morton, W., Goodman, M. G.: Transients and Oscillations in Heterogeneous Catalysis,
1981-2   In: Modelling of Chemical Reaction Systems, K. H. Ebert, P. Deuflhard, and W. Jager
            (Eds.). Springer Ser. Chem. Phys. 18, 253–260

(IIIC)   Nagashima, H.: Chaotic States in the Belousov-Zhabotinskii Reaction. J. Phys. Soc.
1980   Japan 49, 2477–2478

(IIIC)   Nagashima, H.: Experiment on Chaotic Responses of a Forced Belousov-Zhabotinskii
1982   Reaction. J. Phys. Soc. Japan 51, 21–22

(IIIK, C, K) Nair, P. K. R., Mittal, A., Srinivasulu, K.: Chemical Oscillations in the Hydro-
1980-1   quinone-Bromate-Sulfuric Acid System and their Regeneration in the Presence of a
            Catalyst. Izv. Akad. Nauk SSSR, Ser. Khim. (11) 1620–2623 (Rus.)

(IIIK, C, K) Nair, P. K. R., Mittal, A., Srinivasulu, K.: Chemical Oscillatory Reactions with and
1980-2   without Chemical Catalyst. Z. Phys. Chem. (Leipzig) 261(4) 799–801

(IIIK, C, K) Nair, P. K. R., Mittal, A., Srinivasulu, K.: Uncatalyzed Chemical Oscillations in
1981-1   Acidic Bromate Oxidation of Tiron, Ann. Chim. (Rome) 71, 263–267

(IIIK)   Nair, P. K. R., Mittal, A., Srinivasulu, K.: Uncatalyzed Chemical Oscillator Behavior
1981-2   in the Oxidation of Catechol with Acidic Bromate Solution, (Eng.). Bull. Chem. Soc.
            Japan 54, 317–318

(IIIK)   Nair, P. K. R., Mittal, A., Srinivasulu, K.: Chemical Oscillations in the Uncatalyzed
1981-3   Bromate Oxidation of Hydroquinone and Nitrophenols. Reaction Kinet. Catal. Lett. 16,
            399–402

(IIIK)   Nair, P. K. R., Mittal, A., Kadamne, S., Srinivasulu, K.: Uncatalyzed Chemical Oscilla-
1982   tions in the Aqueous Acidic Bromate Oxidation of Veratraldehyde. React. Kinet. Catal.
            Lett 19(1–2) 201–205

(IIIC, D)   Nakajima, K., Sawada, Y.: Experimental Studies on the Weak Coupling of Oscillatory
1980   Chemical Reaction Systems. J. Chem. Phys. 72(4) 2231–2234

113

(IIIC)   Nakajima, K., Sawada, Y.: Phase Diagram for Two Weakly Coupled Oscillatory Systems.
1981   J. Phys. Soc. Japan 50, 687–695

(IIIN)   Nazarenko, V. G., Sel'kov, E. E.: Self-excited Oscillations in an Open Biochemical
1981   Substrate Inhibited Reaction Interacting with the Enzyme Producing System. Biophysika
            26(3) 428–433 (Rus.) [Biophysics 26(3) 435–441 (Eng. trans.)]

(IIIE)   Niiyama, H., and Suzuki, Y.: Oscillations of Catalytic Activity in Hydrogenation of
1982   Ethylene on Nickel-Alumina. Chem. Eng. Comm. 14, 145–149

(IIIC)   Noszticzius, Z.: On the Role of Bromide Ions in the Belousov-Zhabotinskii Reaction
1981   of Malonic Acid. Acta Chim. Acad. Sci. Hung. 106(4) 347–357 (Eng.)

(IIIC)   Noszticzius, Z., Bodiss, J.: Contribution to the Chemistry of the Belousov-Zhabotinskii
1980   Type Reactions. Ber. Bunsenges. Phys. Chem. 84, 366–369

(IIIC)   Noszticzius, Z., Farkas, H.: An Old Model as a New Idea in the Modelling of the
1981   Oscillating B-Z Reaction, In: Ebert et al. (1981). Springer Ser. Chem. Phys. 18,
            275–281

(IIIC)   Noszticzius, Z., Feller, A.: On the Applicability of the Lotka-Volterra Scheme for
1982   Different Types of the Belousov-Zhabotinskii Reaction. Acta Chim. Acad. Sci. Hung.
            110(3) 261–275 (Eng.)

(IIIM)   Noszticzius, Z., Noszticzius, E., Schelly, Z. A.: On the Use of Ion-Selective Electrodes.
1981   I. Potential Response of the Silver Halide Membrane Electrodes to Hypohalous Acids,
            J. Am. Chem. Soc. 104, 6194–6199

(R)   Noyes, R. M.: Oscillations in Homogeneous Systems. Ber. Bunsenges. Phys. Chem.
1980   84(4) 295–303

(IIIC)   Noyes, R. M.: A Mechanism for Bromate Driven Oscillator. J. Am. Chem. Soc. 102,
1980   4644–4649

(IIIB)   Noyes, R. M., Furrow, S. D.: The Oscillatory Briggs-Rauscher Reaction — 3. A
1982   Skeleton Mechanism for Oscillations. J. Am. Chem. Soc. 104(1) 45–48

(IIIA)   Odutola, J. A., Bohlander, C. A., Noyes, R. M.: Chemical Oscillations and Instabilities,
1982   44. Iodide Ion Measurements on the Oscillatory Iodate-Peroxide System. J. Phys. Chem.
            86(5) 818–824

(IIIL)   Orban, M., Dateo, C., De Kepper, P., Epstein, I. R.: Systematic Design of Chemical
1982-3   Oscillators, Part 11. Chlorite Oscillators: New Experimental Examples. Tristability
            and Preliminary Classification. J. Am. Chem. Soc. 104, 5911–5918

(IIIL)   Orban, M., De Kepper, P., Epstein, I. R., Kustin, K.: Systematic Design of Chemical
1981   Oscillators, Part 4. New Family of Homogeneous Chemical Oscillators-Chlorite-Iodate-
            Substrate. Nature (London) 292, 816–818

(IIIL)   Orban, M., De Kepper, P., Epstein, I. R.: Systematic Design of Chemical Oscillators,
1982-1   Part 77. An Iodine-free Chlorite-based Oscillator. The Chlorite-Thiosulfate Reaction in a
            Continuous Flow Stirred Tank Reactor. J. Phys. Chem., 86, 431–433

(IIIC, D)   Orban, M., De Kepper, P., Epstein, I. R.: Systematic Design of Chemical Oscillators,
1982-2   Part 10. Minimal Bromate Oscillator: Bromate-Bromide-Catalyst. J. Am. Chem. Soc.
            104(9) 2657–2658

(IIIL)   Orban, M., Epstein, I. R.: Oscillations and Bistability in Hydrogen-Platinum-Oxyhalogen
1981   Systems. J. Am. Chem. Soc., 103, 3723–3727

(IIIL)   Orban, M., Epstein, I. R.: Systematic Design of Chemical Oscillators, Part 13:
1982-1   Complex Periodic and Aperiodic Oscillations in the Chloride Thiosulphate Reaction.
            J. Phys. Chem. 86, 3907–3910

(IIIL)   Orban, M., Epstein, I. R.: Bistability in the Oxidation of Iron (II) by Nitric Acid.
1982-2   J. Am. Chem. Soc. 104, 5918–5922

(IIIC)   Patonay, G., Noszticzius, Z.: Effects of Stirring the Belousov-Zabotinskii Reaction:
1981   A Problem in Interpretation. React. Kinet. Catal. Lett. 17, 187–189

(IIIA)   Petrenko, O. E. and Grinchuk, A. V.: Autooscillation Model of the Bray-Liebhafsky
1982   Oscillating Reaction based on the Sharma-Noyes Mechanism. Kinet. Katal. 23(1) 22–25
            (Rus.)

(IIIC)   Pikovskii, A. S.: A Dynamical Model for Periodic and Chaotic Oscillations in the
1981   Belousov-Zhabotinskii Reaction. Phys. Lett. 85A, 13–16

(IIIC)   Pomeau, Y., Roux, J. C., Rossi, A., Bachelart, S., Vidal, C.: Intermittent Behavior in
1981   the Belousov-Zabotinskii Reaction. J. Phys. (Paris) Lett. 42(3) 271–273

(IIIN)   Prokopakis, G. J., Seider, W. D.: Adaptive Semi-implicit Runge-Kutta Method for Solu-
1981   tion of Stiff Ordinary Differential Equations. Ind. Eng. Chem. Fundam. 20(3) 255–266

(IIIE)   Rajagopalan, K., Sheintuch, M., Luss, D.: Oscillatory States and Slow Activity Changes
1980   During the Oxidation of Hydrogen by Palladium. Chem. Eng. Commun. 7(6) 335–343

(IIIE)   Rajagopalan, K.: Oscillations in the Rate of Oxidation Reactions Catalyzed by Noble
1981   Metals. U. Microfilms, Order 8202104, Diss. Abstr. Int. B 42(8) 3348

(IIIC)   Ramanathan, S., Ramaswamy, R.: Oscillations in Acid-Bromate-Ferroin System with new
1981   Organic Substrates, Trans. SAEST. 16(4), 241–247

(IIIC)   Ramaswamy, R., Jaya, S., Ganapathisubramanian, N.: Oscillations in the Bromate
1980   System with Malic Acid as the Substrate. Proc. — Indian Acad. Sci.(Ser.): Chem. Sci.
            89, 65–68

(IIIC)   Rastogi, R. P., Rastogi, P.: Oscillatory Chemical Reaction. Part V. Manganese Ion
1980   Catalyzed Belousov-Zhabotinskii Reaction with Acetylacetone. Indian J. Chem. 19A,
            1–6

(IIIE)   Rathousky, J., Hlavacek, V.: Theoretical Investigation of Complex Dynamic Behavior
1981   of the Carbon Monoxide Oxidation on a Platinum Catalyst. J. Chem. Phys. 75,
            749–756

(IIIE)   Rathousky, J., Hlavacek, V.: Oscillatory Bahavior of Long and Short Isothermal Beds
1982   Packed with Platinum/Alumina Catalyst. J. Catal. 75, 122–133

(IIIE)   Rathousky, J., Puszynski, J., Hlavacek, V.: Experimental Observation of Chaotic Behavior
1980   in Carbon Monoxide Oxidation in Lumped and Distributed Catalytic Systems. Z. Natur-
            forsch. 35A(11) 1238–1244

(IIIF)   Richter, P. H., Ross, J.: Oscillations and Efficiency in Glycolysis. Biophys. Chem. 12,
1980   12, 285–297

(IIIF)   Richter, P. H., Ross, J.: Concentration Oscillations and Efficiency: Glycolysis. Science
1981   211(4403) 715–717

(IIIF)   Richter, P. H., Rehmus, P., Ross, J.: Control and Dissipation in Oscillatory Chemical
1981   Engines, Prog. Theor. Phys. 66, 385–405

(R)   Ruoff, P.: Kjemiske Oscillasjoner, Kjemi 6, 26–27 (Norwegian)
1981

(IIIC)   Sakanoue, S., Endo, M.: The Existence of an Unstable Limit Cycle in the Oregonator Model
1982   for the Belousov-Zhabotinskii. Reaction, Bull. Chem. Soc. Japan 55, 1406–1409

(IIIE)   Sales, B. C., Turner, J. E., Maple, M. B.: The Oxidation and Carbon Monoxide Reduction
1981   Kinetics of a Platinum Surface. Surf. Sci. 112, 272–280

(IIIE)   Sales, B. C., Turner, J. E., Maple, M. B.: Oscillatory Oxidation of Carbon Monoxide
1982   over Platinum, Palladium and Iridium Catalysts: Theory. Surf. Sci. 114, 381–394

(IIIE)   Sault, A. G., Masel, R. I.: The Effect of Surface Protrusions on Self-Sustained Thermal
1982   Oscillations during Hydrogen Oxidation on a Nickel Foil. J. Catal. 73, 294–308

(IIIC)   Schlueter, A., Weiss, A.: Nuclear Magnetic Relaxation as Indicator in Oscillating
1981   Chemical Reactions. Ber. Bunsenges. Phys. Chem. 85, 306–309

(IIIE)   Schmitz, R. A., Renola, G. T., Zioudas, A. P.: Strange Oscillations in Chemical
1980   Reactions. Observations and Models. Publ. Math. Res. Cent. Univ. Wis. Madison, 44,
            (Dynamical Model. — Reaction Systems) 177–193

Demet Gurel and Okan Gurel

(IIIN)   Selegny, E., Vincent, J. C.: Model of Enzymic Temporal Chemical Oscillations in Homo-
1980   genous Phase. I. Analytical Study. (Fr.). J. Chim. Phys. Phys.-Chim. Biol. 77,
           1083–1091

(IIIN)   Sel'kov, E. E., Nazarenko, V. G.: Oscillations and Resonance Phenomena in a Simple
                                       E
1981   Enzymic Reaction, → S → P → Interacting with an Enzyme-Producing System. Biofizika.
           26(1) 17–21 (Rus.). [Biophysics 26(1) 13–17 (Eng. trans.)]

(R)   Seno, M., Iwamoto, K.: Theories for Dissipative Structures under Nonequilibrium
1981   States Kazaku (Kyoto) 36, 228–231

(IIIC)   Sevcik, P., Adamcikova, L. U.: Oscillating Heterogeneous Reaction with Oxalic Acid.
1982   Collect. Czech. Chem. Commun. 47, 891–898 (Eng.)

(IIIE)   Sheintuch, M.: Asymmetric Surface States: A Source for Surface Reconstruction and
1981   Structure Sensitivity. Chem. Eng. Sci. 36(5) 893–900

(IIIE)   Sheintuch, M.: Size Dependent Kinetics in a Homogeneously Exposed Catalyst. Chem.
1982   Eng. Sci. 37(4) 591–599

(IIIE)   Sheintuch, M., Luss, D.: Reaction Rate Oscillations During Propylene Oxidation on
1981   Platinum. J. Catal. 68, 245–248

(IIIE)   Sheintuch, M., Pismen, L. M.: Inhomogeneities and Surface Structures in Oscillatory
1981   Catalytic Kinetics. Chem. Eng. Sci. 36(3) 489–497

(R)   Slin'ko, M. G., Slin'ko, M. M.: Autooscillations in the Rate of Heterogeneous Catalytic
1980   Reactions. Usp. Khim. 49, 561–587 (Rus.). [Russian Chemical Reviews 49, 295–309
           (Eng. trans.)]

(IIIM)   Slobodyanikova, L. S., Latov, V. K., Paskonova, E. A., Vitt, S. V., Belikov, V. M.:
1980   The Oscillatory Nature of Enzymic Hydrolysis of Peptide Bonds. J. Mol. Catal. 9(4)
           435–444

(IIIE)   Suhl, H.: Two-Oxidation-State Theory of Catalysed Carbon Dioxide Generation.
1981   Surface Sci. 107, 88–100

(IIIE)   Takoudis, C. G., Schmidt, L. R., Aris, R.: Isothermal Sustained Oscillations in a Very
1981   Simple Surface Reaction, Surf. Sci. 105(1), 325–333

(IIIE)   Takoudis, C. G., Schmidt, L. R., Aris, R.: Isothermal Oscillations in Surface Reactions
1982   with Coverage Independent Parameters. Chem. Eng. Sci. 37(1) 69–76

(R)   Taranenko, A. M.: Dynamic Significance of Substrate Deposition in the Self-oscillating
1980   Biochemical Reactions. I. Cascades of Limit Cycles. Deposited Document VINITI
           3889–80 50 pp. (Rus.)

(IIIN)   Taranenko, A. M.: Sequences of Limit Cycles in a Model of a Biochemical Oscillator
1981   with Substrate Depot. Stud. Biophys. 83, 19–26 (Eng.)

(IIIF)   Termonia, Y., Ross, J.: Oscillations and Control Features in Glycolysis: Numerical
1981-1   Analysis of Comprehensive Model. Proc. Natl. Acad. Sci. U.S.A. 78, 2952–2956

(IIIF)   Termonia, Y., Ross, J.: Oscillations and Control Features in Glycolysis: Analysis of
1981-2   Resonance Effects. Proc. Natl. Acad. Sci. U.S.A. 78, 3563–3568

(IIIF)   Termonia, Y., Ross, J.: Dissipation in an Oscillatory Reaction Mechanism with
1981-3   Periodic Input. J. Chem. Phys. 74, 2339–2345

(IIIC)   Tikhonova, L. P., Zayats, V. Ya.: Study of Oscillating Chemical Reactions with Catalyst,
1980   Complex Ru(II, III) with α,α-dipyridil. Teor. Eks. Khim. 16, 546–551 (Rus.)

(IIIC)   Tikhonova, L. P., Kovalenko, A. S., Yatsimirskii, K. B.: Study of the Joint Action of
1981   Cerium (III, IV) and Manganese (II, III) ions in the Oscillating Belousov-Zhabotinskii
           Reaction Teor. Eks. Khim, 17, 348–355 (Rus.). [Teor. Exper. Chem. 17(3) 267–273 (Eng.
           trans.)]

(IIIM)   Toby, S., Ulrich, E.: Reaction of Carbon Monoxide with Ozon: Kinetics and Chemi-
1980   luminesce. Int. J. Chem. Kinet. 12, 535–546

116

(IIIM)   Tomashov, N. D., Ustinskii, E. P., Chernova, G. P.: Study of Periodic Potential
1981   Oscillations of Chromium Base Alloys with Platinum Group Metals in Sulfuric Acid
            Solution Electrokhimiya 17, 969–976 (Rus.)

(IIIC)   Tomita, K., Tsuda, I.: Chaos in the Belousov-Zhabotinskii Reaction in a Flow System.
1979-1   Phys. Lett. 71A, 489–492

(IIIC)   Tomita, K., Tsuda, I.: Chaotic Behavior in Chemical Reaction Systems. Bussei Kenkyu,
1979-2   33(1) 1–22 (Japanese)

(IIIC)   Tomita, K., Tsuda, I.: Towards the Interpretation of Hudson's Experiments on the
1980   Belousov-Zhabotinskii Reaction (Chaos due to Delocalization). Progress in Theor. Phys.
            64(4) 1138–1160

(IIIF)   Tomita, K., Daido, H.: Possibility of Chaotic Behavior and Multibasins in Forced
1980   Glycolytic Oscillations. Phys. Lett. 79A, 133–7

(IIIC)   Treindl, L. U., Dorovky, V.: A Chemical Oscillator of the Belousov-Zhabotinskii Type
1981   Involving α-ketoglutaric acid. Z. Phys. Chem. 126, 129–131 (Eng.)

(IIIC)   Treindl, L. U., Dorovsky, V.: Kinetics of Oxidation of α-ketoglutaric Acid in Cerium(IV)
1982   Ions in Relation to the Bnlousov-Zhabotinskii Reaction, Collect. Czech. Chem. Commun.
            47, 2831–2837

(IIIC)   Treindl, L. U., Fabian, P.: Influence of Oxygen on the Belousov-Zhabotinskii Reaction.
1980   Collect. Czech. Chem. Commun. 45, 1168–1172

(IIIC)   Treindl, L., Kaplan, P.: Kinetics of Oxidation of 2,4-Pentanedione with Cerium(IV)
1981   Ions in Relation to the Belousov-Zhabotinskii Reaction. Collect. Czech. Chem. Commun.
            46, 1734–1739

(R)   Tsuda, I.: Chaos of Chemical Reactions and Related Problems. Bussei Kenkyu 35,
1981   257–300 (Japanese)

(IIIC)   Turner, J. S., Roux, J. C., McCormick, W. D., Swinney, H. L.: Alternating Periodic and
1981-1   Chaotic Regimes in a Chemical Reaction Experiment and Theory. Phys. Lett. 85A,
            9–12

(IIIE)   Turner, J. E., Sales, B. C., Maple, M. B.: Oscillatory Oxidation of Carbon Monoxide
1981-2   over Palladium and Iridium Catalysts. Surf. Sci. 10, 591–604

(IIIE)   Ukharskii, A. A., Slin'ko, M. M., Berman, A. D., Krylov, O. V.: Periodic Changes in
1981   Concentrations of Surface Compounds during Autooscillations of the Reaction Rate of
            Cyclohexane Oxidation on KY Zeolite. Kinet. Katal. 22(5) 1353–1354 (Rus.)

(IIIE)   Vayenas, C. G., Lee, B., Michaels, J.: Kinetics, Limit Cycles and Mechanism of the
1980   Ethylene Oxidation on Platinum. J. Catal. 66, 36–48

(IIIE, D)   Vayenas, C. G., Georgakis, C., Michaels, J., Tormo, J.: The Role of Platinum Oxide
1981   (PtOx) in the Isothermal Rate Oscillations of Ethylene Oxidation on Platinum. J. Catal.
            67, 348–361

(IIIE)   Vayenas, C. G., Georgakis, C., Michaels, J., Tormo, J.: Response to "Comments on the
1982   Model for Isothermal Oscillations of Ethylene Oxidation on Platinum. J. Catal. 73,
            201–204

(IIIA)   Veljkovic-Slobodanka, R.: Heterogeneity and Rate Modifications of the Bray-Liebhafsky
1981   Reaction. Glas. Hem. Drus., Beograd, 46(11) 711–714 (Eng.)

(IIIC, D)   Vidal, C., Bachelart, S., Rossi, A.: Bifurcations in Turbulent Cascades in the Belousov-
1982   Zhabotinskii Reaction. J. Phys. 43(1) 7–14 (Fr.)

(IIIC, D)   Vidal, C., Noyau, A.: Some Differences Between Thermokinetic and Chemical Oscillating
1980   Reactions. J. Am. Chem. Soc. 102, 6666–6671

(B)   Vidal, C., Pacautl, A. (Eds.): Nonlinear Phenomeny in Chemical Dynamics. Springer
1981   Series in Synergetics, v. 12

(IIIC, D)   Vidal, C., Roux, J. C., Rossi, A.: Quantitative Measurements of Intermediate Species in
1980   Sustained Belousov-Zhabotinskii Oscillations. J. Amer. Chem. Soc. 102, 1241–1245

(IIIC, D)   Vidal, C., Roux, J. C., Bachelart, S., Rossi, A.: Experimental Study of the Transition to
1981   Turbulence in the Belousov-Zhabotinskii Reaction. In: Nonlinear Dynamics (R. H. G.
            Helleman, Ed.). Annals N.Y. Acad. Sci. 357, 377–396

(IIIE)   Wicke, E., Kummann, P., Keil, W., Schiefler, J.: Unstable and Oscillatory Behavior in
1980   Heterogeneous Catalysis. Ber. Bunsenges. Phys. Chem. 84(4) 315–323

(IIID)   Wirges, H. P.: Experimental Study of Self-sustained Oscillations in a Stirred Tank Reactor.
1980   Chem. Eng. Sci. 35(10) 2141–2146

(IIIC)   Yatsimirskii, K. B., Kovalenko, A. S., Tikhonova, L. P.: Special Features of the Action of
1981-1   Two Catalysts in the Belousov-Zhabotinskii Oscillatory Chemical Reaction Dokl. Akad.
            Nauk SSSR 258. 918–923. (Rus.). [Doklady Chem. 258(4) 260–264 (Eng. trans.)]

(IIIC)   Yatsimirskii, K. B., Tikhonova, L. P., Zakrevskaya, L. N., Lampeka, Ya. D.: Homo-
1981-2   geneous Oscillatory Reactions Involving Bromate and Tetraazamacrocyclic Complexes
            of Nickel(II) and (III). Dokl. Akad. Nauk SSSR 261, 647–649 (Chem). [Doklady Chem.
            261(3) 514–516 (Eng. trans.)]

(IIIC)   Yatsimirskii, K. B., Zakrevskaya, L. N., Kol'chinskii, A. G., Tikhonova, L. P.: New
1980   Oscillating Chemical Reactions Involving Macrocyclic Copper Complexes. Dokl. Akad.
            Nauk SSSR 251, 132–134. (Rus.). [Doklady Chem. 251(1) 122–124 (Eng. trans.)]

(IIIC)   Yoshida, T., Ushiki, Y.: The Kinetics of Bromate-Cerium(III) and -Iron(II) Reactions,
1982   Bull. Chem. Soc. Japan. 55, 1772–1776

(IIIC)   Yoshikawa, K.: Distinct Activation Energies for Temporal and Spatial Oscillations in
1982   the Belousov-Zhabotinskii Reaction. Bull. Chem. Soc. Japan 55, 2042–2045

(III)   Young, J., Franzus, B., Huang, T. T. S.: Oscillatory Behavior During the Oxygen
1982   Oxidation of Ascorbic Acid. Int. J. Chem. Kinet. 14, 749–759

(R)   Zhabotinsky, A. M.: Oscillating Bromate Oxidative Reactions. Ber. Bunsenges. Phys.
1980   Chem. 84(4) 303–308

(R)   Zhabotinskii, A. M.: Oscillating Chemical Reactions. Periodic Reactions Based on
1982   Oxidation by Bromate. Kem. Kozl. 57, 23–35 (Hung.)

(IIIC)   Zhabotinskii, A. M., Rovinskii, A. B.: Mechanism of Oscillating Reactions of Bromate
1980   Oxidation, Teor. Eksp. Khim. 16(3) 386–390 (Rus.). [Teor. Exper. Chem. 16(3) 300–303
            (Eng. trans.)]

(IIIC)   Zhabotinskii, A. M., Zaikin, A. N., Rovinskii, A. B.: Auto-Oscillations in the Oxidation
1982   of Trivalent Cerium by Bromate with a Controlled Supply of Bromide into the Reactor.
            Teor. Eksp. Khim. 18(2) 161–165 (Rus.). [Teor. Exper. Chem. 18(1) 137–141 (Eng.
            trans.)]

(IIIE)   Zhang, S. X. M.: Dynamics of Carbon Monoxide Oxidation on Platinum. Diss. Abstracts
1980   Int. B 41(11) 4200

(IIIE)   Zioudas, A. P.: Complex Periodic and Chaotic States in Heterogeneous Catalytic
1980   Systems. Diss. Abstracts Int. B 41(11) 4200

(IIIA)   Zueva, T. S., Protopopov, E. V.: Effect of Anions on Periodic Conditions in the
1982-1   Potassium Iodate-Hydrogen Peroxide-Cysteine System in an Acidic Medium, Teor. Eksp.
            Khim. 18, 364–367 (Rus.)

(IIIA)   Zueva, T. S., Protopopov, E. V.: Study of Oscillating Conditions in the Potassium
1982-2   Iodate-Hydrogen Peroxide-Cysteine System in a Sulfuric Acid Medium. Izv. Vyssh.
            Uchebn. Zaved-Khim. Technol. 25(1) 8–10 (Rus.)

(IIIC)   Zueva, T. S., Sipershtein, I. N.: Study of Oscillatory Conditions in a Citric Acid-
1980   Potassium Bromate-Cerium(IV) Sulfate System in a Sulfuric Acid Medium. Teor. Eksp.
            Khim. 16, 551–554 (Rus.)

(IIIC)   Zueva, T. S., Sipershtein, I. N.: Effect of sec-Butyl alcohol on an Oscillatory Belouzov-
1981   Zhabotinskii Reaction. Teor. Eksp. Khim. 17(2) 277–282 (Rus.). [Teor. Exper. Chem.
            17(2) 217–220 (Eng. trans.)]

Author Index Volumes 101–118

The volume numbers are printed in italics

A. F. Williams

A Theoretical Approach to Inorganic Chemistry

1979. 144 figures, 17 tables. XII, 316 pages
ISBN 3-540-09073-8

Contents: Quantum Mechanics and Atomic Theory. – Simple Molecular Orbital Theory. – Structural Applications of Molecular Orbital Theory. – Electronic Spectra and Magnetic Properties of Inorganic Compounds. – Alternative Methods and Concepts. – Mechanism and Reactivity. – Descriptive Chemistry. – Physical and Spectroscopic Methods. – Appendices. – Subject Index.

This book is intended to outline the application of simple quantum mechanics to the study of inorganic chemistry, and to show its potential for systematizing and understanding the structure, physical properties, and reactivities of inorganic compounds. The considerable development of inorganic chemistry in recent years necessitates the establishment of a theoretical framework if the student is to acquire a sound knowledge of the subject. An effort has been made to cover a wide range of subjects, and to encourage the reader to think of further extensions of the theories discussed. The importance of the critical application of theory is emphasized, and, although the book is concerned chiefly with molecular orbital theory, other approaches are discussed. The book is intended for students in the latter half of their undergraduate studies. (235 references)

Springer-Verlag
Berlin
Heidelberg
New York
Tokyo

Inorganic Chemistry Concepts

Editors: M. Becke, C. K. Jørgensen, M. F. Lappert,
S. J. Lippard, J. L. Margrave, K. Niedenzu, R. W. Parry,
H. Yamatera

Springer-Verlag
Berlin
Heidelberg
New York
Tokyo